MAN and SHELLS

Molluscs in the History

Riccardo Cattaneo-Vietti

Department of Life and Environmental Sciences
Università Politecnica delle Marche (Ancona, Italy)

with Egidio Trainito and Mauro Doneddu

MAN and SHELLS: Molluscs in the History

Author: Riccardo Cattaneo-Vietti

ISBN (eBook): 978-1-68108-225-7

ISBN (Print): 978-1-68108-226-4

First published in 2016.

Acknowledgments:

A special thanks to Nathalie Yonow (Swansea) and Angelo Mojetta (Milan): without the benefits of their experience and expertise, this book would be im- poverished both in scope and interpretation.

We are grateful to Drosos Koutsoubas (University of the Aegean, Mytilene, Gree- ce), Simon Thrush (NIWA, New Zealand), Ester Cecere (CNR, Istituto Talasso- grafico, Taranto), Carla del Vais (Museo Civico di Cabras, Oristano, Sardinia), Salvatore Sebis (Oristano, Sardinia), Giorgio Bavestrello, Maria Cristina Bonci, Antonio Guerci, Paolo Modenesi, Paolo Povero (University of Genoa), Renata Manconi (University of Sassari), Simone Bava (MPA Bergeggi), Vanessa Bracali (Vigo, Spain), Marino Vacchi (ISPRA, Rome), Ferdinando Boero (University of Salento), Gian Carlo Carrada, Giovanni Fulvio Russo, Roberto Sandulli (University of Naples), Lorenzo Senes (Aquarium of Genoa), Carlo Ossola (Milan), Gio- vanni Battista and Antonio Figari (Genoa), Giovanna De Rege (Rocca Grimalda, Alessandria), Giulio Melegari (Milan), for their indispensable suggestions and continued support during the writing.

BENTHAM SCIENCE Bentham e Books

She sells sea-shells on the sea-shore,
For if she sells sea-shells on the sea-shore
Then I'm sure she sells sea-shore shells.

Terry Sullivan, *Tongue-twister (1908) inspired by Mary Anning (1799-1847), the most famous English fossil "hunter"*

Qui zase Bernardin de Ca' Donao
Che morì in tel pescar cape de deo,
co la camisa curta e l'cul bagnato
Del Milecinquecento. Ora pro eo.

(Here Bernardin de Ca' Donao lies. He died fishing scallops with a short shirt and a wet ass. In the 1500 year. Pray for him)

A scallop fisherman's epitaph in Venice (16th century)

TO GIANNI ROGHI

This book is dedicated to the memory of Gianni Roghi (1927-1967), Italian journalist, diver, photographer, mollusc expert, and curious explorer of the world.

Keen collector of shells, Roghi left us from the pages of *Mondo Sommerso*, the first world magazine dedicated to diving activities, a series of articles on malacology which were, for the youth of those times, a treasure of knowledge, research, and inspiration.

Cover: a painted *Nautilus* shell from Bali, Indonesia (20th century).

CONTENTS

Authors

Riccardo Cattaneo-Vietti

Full Professor in Ecology at the Department of Life and Environmental Sciences of the Università Politecnica delle Marche (Ancona), he was born in Genoa (Italy) on 16th July 1949. Graduated in Biological Sciences at the University of Genoa (1972), he is interested in marine ecology and biology, in particular the study of Mediterranean and Antarctic benthic communities. He is a specialist on the opisthobranchs molluscs and has been responsible for several research projects related to benthic communities and the management of marine protected areas (Sistema Afrodite, Fish and Spill-over, ASPIM I and ASPIM II).
Regional Editor of *Marine Biology* for a decade (1997-2007), he is now Associate Editor of *Aquatic Biology* journal.
Author of 9 books on marine ecology topics and fisheries, he summarized his studies in over 380 scientific publications, mostly published in international journals. His interest were focused on structure and dynamics of littoral environments, such as cave and coralligenous communities, on the red coral biology, the effects of sedimentation on the benthic communities, and the influence of mineralogical features of substrate on settling and community structure.
In Antarctica, he worked on population dynamics, reproduction, trophic relationships, and benthic-pelagic coupling.
He represented the University of Genoa in the Board of the Marine Protected Area of Portofino Promontory and is a member of the Scientific Council of the Marine Protected Area of Tavolara - Capo Coda Cavallo. He has been a member of the scientific board of the Italian Oceanographic Society and member of several sci-

entific associations (Italian Marine Biology Society, Italian Malacological Society, and Unitas Malacologica).
He has taken part in several scientific expeditions around the world (Norway, Scotland, Capo Verde Islands, Chile, Antarctica, Great Barrier Reef, Indonesia, Japan).
He served on the Board of the National Consortium for Marine Sciences (CONISMA) 2003-2012 and was Director of the Marine Biology Center at the University of Genoa. In 2010, he received the Golden Trident Award from the International Academy of Underwater Sciences and Techniques.

Mauro Doneddu

Doctor in Medicine, he is actively interested in various fields of zoological, botanical and environmental sciences. Since 1980, he is a member of the Italian Society of Malacology. He published dozens of scientific publications in malacology on national and international journals. He published also several monographs, including: *Guide to the butterflies and orchids in Sardinia* and *Guide to wild species*, in collaboration with other authors. It is co-author of the *Guide of the Mediterranean Shells*, now in its second edition.

Egidio Trainito

Egidio Trainito lives in Sardinia: nature photographer and consultant, he is mainly involved in national and international projects for sustainable development of tourism and for conservation in Protected Areas.
As a free lance, he has traveled the seas of the world and has put more than 4000 dives on his weight belt.
Member of SIBM (Italian Society for Marine Biology), he is the author of many books on marine environments and Marine Protected Areas.
For Il Castello Editore, he is the curator of a series whose most important titles are *Atlante di Flora e Fauna del Mediterraneo*, *Nudibranchi del Mediterraneo* and *Spugne del Mediterraneo*.
Since 2005, he is author/consultant of the TV series *Lineablu* RAI 1.

Foreword by Nathalie Yonow

Man has always had a close association with molluscs and their shells, be it for nutrition or inspiration, for art and culture, or for more practical purposes such as bartering, currency, and medicine. This association goes back to the Palaeolithic, the beginnings of Man: snails and bivalves were good renewable sources of protein, and their shells could be utilised in many ways: early Man wasted nothing. Over time, these shells not only played an important role in the material survival of Man, but also allowed him to develop his culture, with symbolic values of great importance. At the dawn of history, shells assumed mythical meanings and rituals, objects of life and death, of high symbolic value, status, and magic.

Molluscs produce wonderful shells, and these shells display shapes and designs that have been and still are objects of astonishment for all, and sources of inspiration for artists of all epochs. The sciences, medicine, religion, literature, jewellery, world trade, and gastronomy all owe much to these producers of shells. They include species very familiar to us, such as garden snails and nautilus, mussels and oysters, limpets and clams, and so many others. The contribution that molluscs have made, and continue to make to both experimental biology and medicine is immense. Gastropods and squids are still used as lab animals: they seem to have been designed by Nature itself to facilitate research on neurology and biochemistry. Some species even produce chemicals which may be used by pharmacology in the cure of disease, and the shells themselves are currently being investigated for the repair of damaged cells causing paralysis (Centre for Paralysis Research at the Purdue School of Veterinary Medicine, Indiana, USA). The shells of molluscs from all over the world, land, lakes, and ocean, contain very detailed imprints of past climate change and with isotope analysis, these signals have been extracted to demonstrate long-term climate variations.

This book is not simply a treatise of malacology: it is a tribute to a group of organisms that has always fascinated and inspired Man, accompanying him throughout his history, a group of organisms that has had and continues to play an important role in the cultural evolution of Man. The book is not full of scientific detail, but instead full of prose and poetry, stories and anecdotes based on historical facts, beautifully illustrated with images of ancient and modern artefacts celebrating the amazing world of shells.

Swansea University, Wales, UK

Foreword by Angelo Mojetta

A first thought springs to mind after reading this essay by Riccardo Cattaneo-Vietti: the molluscs have finally been vindicated, all the attention they deserve, has been obtained. Although often poorly evaluated, there is no other group of organisms which has interacted with Man since the dawn of civilization, and in such an extensive and widespread manner. Flipping, even summarily, through these pages, you will realize that molluscs have accompanied humanity in the course of his evolution, and perhaps even saved a part of it from the hunger, when their edibility was discovered and appreciated. Honour and glory to shellfish, then.
A second thought: molluscs can stimulate all the five human senses: a very rare, if not unique peculiarity. Does this seem a risky assertion? Absolutely not: try thinking of your personal experience, when some molluscs crossed your path, and you will discover that they have interacted with you or you with them more often than you think, and always through one or more of your senses. The first contact with shellfish is certainly through taste. To find this out, just browse a cookbook or remember what was the last shellfish or squid you ate. Closely related to taste, then, is the sense of smell. It is a sore point. More than the taste, molluscs can stimulate our dis-taste. If fresh or well-cooked shellfish emit pleasant aromas, very different from each other and difficult to describe: the iodized salty oysters, the abalones softness, the scallop delicacy, the fleshiness of the octopus, the toughness of the muricids. Big differences in flavours, that make each species unique. Not exactly the same occurs when their meat begins to decompose and putrefy. Try to forget some mussels for few days in your car and then you will understand what I mean! However, the mollusc, before being eaten and appreciated, must be collected: so sight is involved, a sense really stimulated by shellfish. Watch a shell, not just to recognize it, but to admire its shape and infinite varieties of colours. This is, and has been, the delight of generations of collectors, whose age ranges from five to one hundred. But the form also urges the touch, and there is no doubt that to hold a pearl, a cowrie, or a *Nautilus* shell can be the source of great enjoyment for everyone. If you have read this far, you will note that there is still a last sense: hearing. To find out in detail about this, just read the Chapter 3, but maybe you just have think back to the last time you listened to the murmurs of the sea imprisoned in the shell whorl, holding its opening close to your ear. It is true that the science explains that is not the sound of the sea, but it is so much more fun to believe it.It would be a shame to surrender to physical law, avoiding the chance to dream. Molluscs, apparently, also serve this purpose.

Marine biologist and journalist, Italy

PREFACE

All the facts in natural history, taken by themselves, have no value, but are barren like a single sex. But marry it to human history, and is full of life.

Ralph Waldo Emerson (1803-82), *Address on Nature* (1836)

From the Paleolithic age to present, Molluscs have always played a role, often unknown, in History.
Their shells show shapes and designs that have been and are still object of astonishment for all and source of inspiration for the artists of all time. But not only: sciences, medicine, religion, literature, art, jewellery, world trades, and, of course, gastronomy, owe much to these producers of shells. They include species very familiar to us, such as garden snails and octopuses, mussels and oysters, clams and squids, others strange and unknown.
Today, the shellfish are major food resources for many populations: their demand from the international market is always in growth: they are also intensively farmed both at sea and on land, taking into account that for some "wild" populations, the threat of overexploitation is real.
Over time, the shells did not only play an important role in the material survival of Man, but also allowed to develop his culture, taking on symbolic values of great importance. At the dawn of history, shells assumed mythical meanings and rituals too: objects for life and death, of high symbolic value and magical: amulets, sexual symbols, ornaments of beauty.
Shells and Molluscs have been also a constant source of inspiration for artists: their geometries inspired many architects, while an incredible number of painters and sculptors of all time have copied their shapes. Since the Renaissance, the shells are very popular, objects of desire for many, and now a hobby involving tens of thousands of fans around the world. The shells often show such refined beauty to arouse a sense of pleasure, an aesthetic pleasure in the myriad of passionate collectors who, in all times and in all over the world, always looking for and preserved, sometimes in almost pathological way.
Sometimes, they were used as musical instruments: rattles and trumpets made with shells accompanied the children to play and the rites of war and peace of

many people, in at least four continents, while some cowries were “true” money in three continents for over half a millennium.

Molluscs have had also an important role in the history of clothing: the Phoenicians learned to extract purple to dye clothes from Molluscs and until the last century in southern Italy fine "silky" tissues were made spinning the byssus from a Mediterranean bivalve.

The contribution that Molluscs have given and continue to give to the experimental biology and medicine is great. Gastropods and squids were and are still used as lab animals: they seem to have been designed by Nature itself to facilitate researches on neurology.

This book wants to be just a tribute to a group of organisms that, consumed as food since prehistoric times, has always fascinated, and inspired Man, accompanying him in his history: a group of organisms that has and continues to have an important and little known place in the cultural evolution of Man.

CONFLICT OF INTEREST

No conflict of interest arised in this work.

Riccardo Cattaneo-Vietti
Department of Life and Environmental Sciences
Università Politecnica delle Marche (Ancona, Italy)

ACKNOWLEDGEMENTS

A special thanks to Nathalie Yonow (Swansea) and Angelo Mojetta (Milan): without the benefits of their experience and expertise, this book would be impoverished both in scope and interpretation.

We are grateful to Drosos Koutsoubas (University of the Aegean, Mytilene, Greece), Simon Thrush (NIWA, New Zealand), Ester Cecere (CNR, Istituto Talassografico, Taranto), Carla del Vais (Museo Civico di Cabras, Oristano, Sardinia), Salvatore Sebis (Oristano, Sardinia), Giorgio Bavestrello, Maria Cristina Bonci, Antonio Guerci, Paolo Modenesi, Paolo Povero (University of Genoa), Renata Manconi (University of Sassari), Simone Bava (MPA Bergeggi), Vanessa Bracali (Vigo, Spain), Marino Vacchi (ISPRA, Rome), Ferdinando Boero (University of Salento), Gian Carlo Carrada, Giovanni Fulvio Russo, Roberto Sandulli (University of Naples), Lorenzo Senes (Aquarium of Genoa), Carlo Ossola (Milan), Giovanni Battista and Antonio Figari (Genoa), Giovanna De Rege (Rocca Grimalda, Alessandria), Giulio Melegari (Milan), for their indispensable suggestions and continued support during the writing.

MAN and SHELLS

CHAPTER 1

Not only Shells

Abstract: Molluscs show an architecture designed by an inexhaustible imagination, shapes and designs that have been and are still a marvel for all, and a source of inspiration for artists of all time. But not only that: science, literature, jewellery, handicraft, and, of course, gastronomy owe much to the molluscs, the shells' producers. They include forms very familiar to us, such as the garden snails and the octopuses, mussels and oysters, clams and squid and appeared on our planet long before Man, approximately 500 million years ago.
At some point, they crossed the Man, playing an important role in our life. Many molluscs, in fact, have become part of the human diet from the beginning of our cultural evolution, while others become natural containers, and their uses were innumerable. Some shells have dominated among the talismans for centuries, sometimes as special witness of religious pilgrimage, or represented money in three continents for more than half a millennium. Moreover, the whelks, the Triton trumpets, can be considered one of the first communication instrument at the global level, while, other molluscs, like the Mediterranean *Bolinus brandaris* and *Pinna nobilis*, producing purple and byssus respectively, have had an important role in the history of clothing.
Finally, the contribution that molluscs have provided and continue to give to the experimental biology as well as in mathematic studies is great and, since the classical ages, they are also part of literature and were, sometimes, protagonists even at cinema and in music.

Keywords: Aliens, byssus, bivalves, cinema, collection, cones, cowries, *Dreissena*, farming, gastronomy, gastropods, literature, molluscs, music, mussels, oysters, purple dye, shell structure, Shell Co, Saint Jacques shell, squids, trades.

Riccardo Cattaneo-Vietti, Egidio Trainito and Mauro Doneddu

On a closed book of famous bards, far away you rest
from mother moss, like a marble rest,
formerly waves' daughter, oh twisted shell.
Hidden in the marine cave darkness,
you saw the morning of young Man's world;
wandered with nautilus, and with muricids you lived;
but Man was not there.

Giacomo Zanella (1820-1888), Italian poet, *About a fossil shell on my desk*

Molluscs and their shells have been part of our daily lives, not just in the kitchen. Once, all of us collected a shell, walking along a beach, perhaps only by holding its opening close to the ear, to hear the murmurs of the sea imprisoned in its whorls or for the giving of a gift, a sign of love, or keeping it as a souvenir of a holiday. Maybe, we did not even realize it, but a shell, among all the natural objects in our homes, is the most common: perhaps forgotten in the bottom of a drawer, but it is there!

This book is not a treatise on Malacology, the science that studies molluscs: it is simply a tribute to a group of organisms that, consumed as food since prehistoric times, has always fascinated and inspired Man, accompanying him in his history. Architecture designed by an inexhaustible imagination, all perfectly adapted to the environmental conditions in which their builders live, shells show shapes and designs that have been and are still a marvel for all, and a source of inspiration for artists of all time. But not only that: science, literature, jewellery, handicraft, and, of course, gastronomy owe much to the molluscs, the shells' producers. They include forms very familiar to us, such as the garden snails and the octopuses, mussels and oysters, clams and squid. As the Italian poet Giacomo Zanella reminds us in his lyric (1864), *About a fossil shell on my desk,* molluscs appeared on our planet long before Man, approximately 500 million years ago, living through a complex biological evolution that, at some point, crossed with our cultural evolution. Taken all together, molluscs are a very heterogeneous group of invertebrates, second only to arthropods (insects and crustaceans), as numbers of species present on Earth: nearly 100,000 are currently known (Haszprunar, 2001), and probably a similar number disappeared, sacrificed to countless evolutionary processes. In 500 million years of evolution, they have been able to conquer all environments on the planet, from sea to freshwater, from mountains to ocean depths.

The molluscs are the second most diversified group on the animal kingdom after arthropods, with some 100.000 living species currently known world-wide. Within this enormous diversity, eight classes are recognized: Gastropoda which include limpets, conchs, whelks, snails and slugs, Bivalvia, with oysters, scallops, clams and mussels, Polyplacophora, the chitons or coat-of-mail shells, Cephalopoda, octopus and squids, Scaphopoda, the tusk shells, while Monoplacophora, Solenogastra, and Caudofoveata are minor classes.

Molluscs are characterized by being invertebrates which have soft, unsegmented bodies that are frequently protected by a calcareous or proteinaceous structure, the shell. Although normally external, this structure can be located internally in some groups. The shell may be formed by a single piece (in gastropods, scaphopods, monoplacophorans, and in cephalopods, if they have one), by two pieces in bivalves, on by eight individual plates in polyplacophorans. In the case of the Solenogastra and Caudofoveata, the surface of the body appears to be covered by small calcareous spicules. The shell may be completely absent in some gastropods (nudibranchs and octopods).

The molluscs' main characteristic is the shell, a defensive structure, often calcified, in which they can retract and protect themselves. It is always present during their larval stage even if lost during the development. Being soft organisms, they are able to produce solid "houses", often very complex in their shape.

Figure 1: A *Turbo sarmaticus* shell. The surface has been slightly polished to highlight the pearly inner layer.

Basically, the shell structure is the result of the deposition of a series of layers secreted by a special organ, the mantle, an epidermal fold which covers the external surface of the body. The outermost layer, *periostracum*, consists of a thin film of a sclerotic protein, called conchiolin, while the others consist of two forms of calcium carbonate, calcite and aragonite, with different degrees and patterns of crystallization (Fig. **1**). The fundamental role of the protein layer is to protect the carbonates from the sea water's corrosive properties and in some species, as the cowries in which the mantle completely surrounds the organism, this layer can also be absent. Among the calcareous layers, the outermost (*ostracum*) is composed by calcite in a prismatic, not iridescent, structure while the inner one (*ipostracum*) is formed by lamellar aragonite layers. If these last layers reach a certain thickness they become iridescent to form the nacre or mother-of-pearl. Not all the shells have a layer of nacre, but some, with the exception of thin layers of *periostracum* and calcite, are almost completely made by nacre. The number of these calcareous layers is variable, and can reach a maximum of six in some gastropods: it is easy to see the different layers, looking carefully at a washed ashore shell, eroded by the waves.

The shells often show very complex designs and bright colours, as a result of the combination of four basic pigments: carotenoids for yellow-orange, indigo for the blue, porphyrins for the green, and melanin for the dark colours. All these pigments are secreted by the mantle glands, and are going to fix into the calcite, before it hardens.

This shell, which during their evolution, reached sophisticated and complex shapes, will be later abandoned in the adult stage by the more evolved molluscs (such as nudibranchs, many pulmonates and almost all cephalopods), replaced by much more sophisticated chemical defence mechanisms. But the shell remains, however, the characterizing feature of this group.

Some molluscs, especially bivalves, are sessile, living anchored to a substrate often using the byssus threads, or burrowed in the sediments. Others, on the contrary, are mobile; these include most of gastropods, and they are often also strong swimmers, such as the cephalopods. Some squids can also “fly” for a short time, covering distances as far as 30 m above the surface of the water thanks to their jet propulsion, while some small marine gastropods (pteropods and heteropods) "fly" in the sea, floating in the water column (Lalli and Gilmer, 1989). Finally, a few bivalves, like scallops, are capable of a limited form of swimming.

Not all molluscs are striking: many are small in size, as *Omalogyra atomus* that rarely exceeds one millimetre (Doneddu and Trainito, 2010) or even smaller, belonging to the interstitial fauna that lives among the sand grains and that certainly escapes the attention of most. But, among them, we find also the largest living invertebrate, *Architeuthis dux,* the giant squid whose stretched tentacles can exceed 17 metres in length (Ellis, 1999). This is the fabulous and terrific kraken, the "monster" that so astonished and frightened the Northern European sailors, when they were pushed to explore the Western Atlantic.

Among bivalves, the south-western Pacific giant clam, *Tridacna gigas,* is the largest living one, reaching a length of 130 centimetres and a weight of more than 250 kilograms. Even the shell of a tropical ship-worm, *Kuphus polythalamia,* is very long, reaching 150 cm in length, but is very thin and weights very little. Sometimes, they are very long-living species: a Pacific geoduck, *Panopea abrupta,* of 7 kg, common in the north-east Pacific could live as long as 150 years, while some deep-water bivalves, such as the “ocean quahogs”, *Arctica islandica*, were dated back 300-400 years. Among the freshwater bivalves, the pearl mussel, *Margaritifera margaritifera,* seems to exceed a century in age. It is obvious that when a species has also a commercial value and is actively harvested (as in all these cases), the risk of quickly destroying its population becomes very high.

Chitons, gastropods, and cephalopods are generally herbivores or carnivores, sometimes scavengers. Few of them, like vermetids, pteropods, and heteropods feed on the organic matter suspended in the water column, using mucous traps. Most bivalves and scaphopods feed, instead, by filtering water, or collecting organic detritus settled on the sea bed. The filtration capacity of a bivalve is considerable: a mussel can reach a capacity of 3 litres per hour (Jorgensen, 1990). This is the reason it is very dangerous to eat filter-feeding bivalves harvested in polluted waters: the shellfish could, in fact, have accumulated in its filter apparatus or in its digestive gland (hepatopancreas) pathogens or toxic substances which, if ingested, can cause serious diseases. A few molluscs are pests, but some of them, especially from freshwater, enter in the biological cycles of many parasites, some of them also of interest for human health. Very few molluscs are poisonous to humans, but someone, having eaten poisonous organisms (usually microalgae, sponges, and cnidarians), can acquire their toxicity and, in turn, become dangerous even on simple physical contact. Others, mainly due to acid secretions, are able to bore through rock or wood, often creating huge damage to man-made artefacts immersed at sea.

Some species, also as result of maritime traffic and trades, have made long trips, going from one ocean to another, from one biogeographical region to another: this is the case of many species which, after the opening of the Suez Canal in 1869, have "invaded" the Mediterranean Sea from the Red Sea. These species are known as "lessepsian" migrants, in memory of Ferdinand de Lesseps (1805-1894), promoter and executor of the Canal. The alien molluscs found in the Mediterranean, lessepsian or not, number approximately 130 (Gofas and Zenetos, 2003) and are still rising: just three years after this publication, they counted nearly 170. Today, these species are reported mainly in the Eastern Mediterranean and Adriatic Sea (Çinar *et al.*, 2005), and some of them have established stable populations, quickly adapting to the new environmental conditions. In the absence of competitors, some have shown a huge quantitative increase, also creating problems by entering into competition with the native populations of similar species, and to human productive activities (**Chapter 12**).

Among the most harmful bivalves, the zebra mussel, *Dreissena polymorpha*, is quite famous: this northern European freshwater species reached the American Great Lakes and the Mississippi basin at the end of the 20th century, probably as a result of the importation of other species of commercial interest. Quickly it started to develop in a huge manner, obstructing pumping water installations in lakes and rivers and causing important economic damage. Of course, many molluscs have become part of the human diet from the beginning of our cultural evolution (Fernàndez-Armesto, 2002): raw or cooked, prepared in many ways, they are a tasty food and a good source of protein. There is no culture that does not list in his kitchen some shellfish and it is very likely that the high numbers of bivalve populations inshore or along rivers and lakes have favoured the changing habits of primitive Man: he moved to a more sedentary way of life, exploiting them very easily to harvest. Today, the shellfish are major food sources for many human populations: their demand in the markets always shows a strong growth: they are also intensively farmed both at sea and on land, taking into account that for some "wild" species, the threat of over-exploitation is real. Among the main marine species of a particular economic interest, fished or farmed around the world, some are, in fact, molluscs: in particular an oyster (*Crassostrea gigas*), a clam (*Ruditapes philippinarum*), the Eastern scallop (*Patinopecten yessoensis*), a mussel (*Mytilus edulis*) and, in addition, two squids, *Illex argentinus* and *Todarodes pacificus*. Some of these, and oysters in the first place, can be considered a real delicacy, revered by gourmets and poets since ancient times. The appreciation for oysters has no regional boundaries or cultural limitations. Perhaps nobody would ever expect that the unfortunate Italian poet Giacomo Leopardi (1798-1837), consumed by bone tuberculosis, could have written, a year before his death:

Qual puoi bramar felicità più vera
che far d'ostriche scempio infra gli amici ?
Sallo Santa Lucia, quando la sera,
poste le mense, al lume delle stelle,
vede accorrer le genti a schiera a schiera,
e di frutta di mare empier la pelle.

(What more real happiness that makes oysters havoc among friends?
We do at St Lucia, when night falls. The tables are laden in the light of the stars, a multitude of people rushes, looking forward to gorging with seafood).

This amazing passion for oysters may have led to the death of the poet, the cause of which was never fully discovered: perhaps an oyster eaten in Naples, during an outbreak of cholera. Always in Naples, another great Italian poet, Giosuè Carducci (1835-1907), wrote: *Every day, I eat twelve oysters, drinking a bottle and half or two of Posillipo or Vesuvius wine*. It seems no coincidence that the poet died of cirrhosis of the liver!

Oysters are considered an aphrodisiac food *par excellence* thanks to Giacomo Casanova (1725-1798) who, in his *Histoire de ma vie*, enhances their properties, by making them a symbol of sensuality. It is not a coincidence that two pornographic magazines during the Victorian age will later be called *The Oyster* and *The Pearl* (Stott, 2004), also recalling the concept of pearl, a symbol of perfection and beauty, closely linked to desire and love. In the aphrodisiac properties of the oysters, however, some truths remain: they contain, in fact, high levels of zinc, an element that would favour the synthesis of testosterone in males (Ebisch *et al.*, 2007).

Many shells, once abandoned by their owners, become natural containers and their uses were innumerable. In nature, snail empty shells are used by a small bee, *Osmia bicolor*, to lay its eggs, but Man began to use them for many uses, adapting them to his needs. Liquid containers, plates, spoons, paint pots, chisels to engrave wood, hooks for fishing. In many different cultures, craft practiced for centuries included beadwork, often made with shells. Today, these objects are shown in ethnological museums but, until a few years ago, many people in Africa, South America, and in the South Seas continued to use similar tools.

Over time, shells do not only play an important role in the material survival of Man, but also allowed him to develop himself, taking on symbolic values of great importance. At the dawn of history, the shell assumed, in fact, a mythical meaning in rituals: in South African deposits, more than 70,000 years ago, many shells were found with bone fragments and splinters of stone, certainly the remains of ancient meals, but also the ritual objects for life and death, with high symbolic and magical values, amulets, sexual symbols, beauty ornaments.

Among shells, some cowries (*Monetaria moneta, M. annulus*) from the Indian Ocean, have dominated among the talismans for centuries, as well as having represented money in three continents for more than half a millennium. In the 19th century, the value of a young wife was 60,000 to 100,000 cowries; an older, more ordinary wife maybe reached 20,000 cowries (Rogers, 1908). The use of cowries as money dates far back into antiquity. They were found in the ruins of Nimrud, and Marco Polo saw them in Yunnan, in the 13th century. In China, shell was and still is the symbol of luck and prosperity, as well as was a symbol of the Emperor: to emphasize that he himself was "good luck and prosperity" for his Empire! From magical rites to religious ones, the step was short: shells have become, over time, important symbolic value in many religions. Hinduism has a sacred shell, the “chank”. In the medieval Christianity, they took different symbolism: the scallop or "Saint Jacques shell" (*Pecten jacobaeus* and *Pecten maximus*), became, over the centuries, a special witness of pilgrimage, while the pearl oyster was one of the symbols of the Mother of God, who holds and protects the pearl, Christ himself. The idea of the shell-container, inside which we can find a valuable asset, is much older: according to the ancient Greek myth, the new-born Aphrodite was lying inside a shell, to protect her.

Sometimes, the shells of large gastropods were also used as musical instruments: rattles and trumpets, made with shells, accompanying children at play, and for rites of war and peace of many people in at least four continents. The whelks, with a broken apex, can be considered the first wind instruments of communication on the sea at the global level. In the Indo-Pacific islands, until a few years ago, the fishermen recalled blowing in a *Charonia tritonis*, exactly like their Mediterranean colleagues with *Charonia lampas* or *C. variegata*. Today, all these species are subject to some form of protection, and their marketing is under control.

Figure 2: A section of an ammonite.

Some molluscs are true living fossils as the *Nautilus* species which belong to a group of cephalopods, closely allied to the ammonites (Fig. **2**), practically extinct during the Paleozoic. The first living specimen was described only in 1832, surprising many palaeontologists. In reality, their white and smooth shell, with orange-red streaks, was already well known in Europe since at least the 16th century, thanks to trade with the East Indies and widely used in jewellery. Today, it is quite common, and protected, in the Indo-Pacific, especially from Fiji Islands to New Guinea.

Molluscs have also had an important role also in the history of clothing: the Phoenicians, in fact, learned to extract the purple to dye tissues from some gastropods, and until the last century in southern Italy, the byssus of the noble pen shell, *Pinna nobilis*, the largest Mediterranean bivalve, was spindled to produce fine "silky" tissues.

A shell, or rather a scallop (Fig. **3**), is the brand of one of the largest oil companies in the world (Cox, 1957). It is worth recalling its story: in 1897, Sir Marcus Samuel (1853-1927) and his brother Samuel (1855-1934) founded the *Shell Transport and Trading Company*, the first European oil company, and chose as symbol of the company a shell in memory of their father who traded with the East in objects decorated with shells or shells themselves since 1833. Being situated on the Black Sea, the Samuel brothers thought that trading oil would be more profitable for business than trade in shells! In 1890-92, they built eight oil tankers (the first was called *Murex*, also a mollusc) to bring Russian kerosene to Japan through the Suez Canal. This was the first business of what was to become one of the most powerful multinational oil companies, born in the back room of a shells merchant!

Another shell has become well known in the world of luxury when, in 1926, Hans Wilsdorf, founder of the *Rolex*, called his first waterproof watch *Oyster*, to emphasize its strength and imperviousness.

Often shells appear in coats-of-arms of some European cities. In England, the County of Whitby coat-of-arms shows ammonites, fossil molluscs very common in that area (**Chapter 2**), while in the United States, Rhode Island adopted the northern quahog (*Mercenaria mercenaria*) (Fig. **46**), the hard-shelled clam and one of the main foods of the local Indians, as symbol of the state, while the knobbed whelk, *Busycon carica*, is the state shell of both New Jersey and Georgia

Figure 3: The scallop shape: brand of one of the largest oil companies in the world.

In Siena (Italy), the coat-of-arms of the Contrada del Nicchio ("nicchio" means shell) shows a shell, surmounted by a grand-ducal crown and surrounded by a coral necklace, roses, and knots of Savoy family, while on the shield of the Contrada della Chiocciola ("chiocciola" means snail) a snail is crawling on a silver lawn. In Bahrain, one of the most historic areas of pearl fishing in the Persian Gulf, a great monument reminds visitors of the importance that the natural pearl fishing had been in the local economy, while on the flag of the ancient kingdom of Travancore, in the Indian state of Kerala, the *chank,* the shell sacred to Hinduism, stands on a red field.

Since the classical ages, the molluscs are also part of literature: Juvenal, Martial and Horace praised the gastronomic qualities of oysters, but it is especially from the 18th century onwards that we meet in world literature more and more references to shellfish, as in *Polite Conservation* by Jonathan Swift (1731), Guy de Maupassant's *Bel Ami* (1885) or Marcel Proust (1871-1922), the refined writer of the Belle Époque in Paris. In *Doktor Faustus* by Thomas Mann (1875-1955), the troubled Jonathan Leverkühn, always hovering between science and faith, observing the complex design of a shell from New Caledonia, wonders whether it hides a mysterious message from Nature, while a trumpet-shell assumed an important symbolic feature in the famous novel by William Gerald Golding (1911-1993), *Lord of the Flies*.

Sometimes, the molluscs have been described as dangerous and terrible monsters: the fight between Gilliat and an octopus in the popular novel *Les Travailleurs de la Mer* by Victor-Marie Hugo (1802-1885) is famous, remembering the ancient myth of Hercules who fights against Hydra, the water-snake with many heads that lived in the Lerna swamps. In the same years, Jules Verne (1828-1905) published *20,000 Leagues under the Sea* (1870) in which molluscs are often protagonists, while the North American poet Oliver Wendell Holmes (1809-1894) wrote a romantic poem, *The Chambered Nautilus*, on the fabulous cephalopod whose shell inspired many artists with the purity of its form.

This is the ship of pearl, which, poets feign,
Sails the unshadowed main,
The venturous bark that flings
On the sweet summer wind its purpled wings

In gulfs enchanted, where the Siren sings,
And coral reefs lie bare;
Where the cold sea-maids rise to sun their streaming hair.
Its webs of living gauze no more unfurl;
Wrecked is the ship of pearl!
And every chambered cell,
Where its dim dreaming life was wont to dwell,
As the frail tenant shaped his growing shell,
Before thee lies revealed.
Its irised ceiling rent, its sunless crypt unsealed!
Year after year beheld the silent toil
That spread his lustrous coil;
Still, as the spiral grew,
He left the past year's dwelling for the new,
Stole with soft step its shining archway through,
Built up its idle door,
Stretched in his last-found home, and knew the old no more.
Thanks for the heavenly message brought by thee,
Child of the wondering sea,
Cast from her lap, forlorn!
From thy dead lips a clearer note is born
Than ever Triton blew from wreathed horn!
While on mine ear it rings,
Through the deep caves of thought I hear a voice that sings:
Build thee more stately mansions, my soul,
As the swift seasons roll!
Leave thy low-vaulted past!
Let each new temple, nobler than the last,
Shut thee from heaven with a dome more vast,
Till thou at length are free,
Leaving thine outgrown shell by life's unresting sea!

Even Lord Alfred Tennyson (1809-1892), the Victorian English poet, fascinated by the giant squids which sometimes were found stranded on the beach, described this mythical mollusc in the poem *The Kraken* (1830) as a monster asleep, intending to re-emerge at the end of the world:

Below the thunders of the upper deep,
Far far beneath in the abysmal sea,
His ancient, dreamless, uninvaded sleep
The Kraken sleepeth: faintest sunlights flee
About his shadowy sides: above him swell
Huge sponges of millennial growth and height;
And far away into the sickly light,
From many a wondrous grot and secret cell
Unnumbered and enormous polypi
Winnow with giant fins the slumbering green.
There hath he lain for ages and will lie
Battening upon huge seaworms in his sleep,
Until the latter fire shall heat the deep;
Then once by men and angels to be seen,
In roaring he shall rise and on the surface die.

Sometimes, molluscs have been used as a benchmark for their peculiar characteristics in moralizing poetry, in the wake of the classical and medieval bestiary fables. Among the various aphorisms and apologists, the famous Roman poet Trilussa (1871-1950) wrote:

La Lumachella de la Vanagloria,
ch'era strisciata sopra un obbelisco,
guardò la bava e disse: Già capisco
che lascerò un'impronta ne la Storia.

(The small snail, full of vainglory,
was crawling over an obelisk.
It looked at its slime and said:
Yeah, I understand I'm going
to leave a trace in History).

A story of presumption and vanity that is still very present today. More recently, the Italian poet Eugenio Montale (1896-1981) titled his most famous collection of poems, *Ossi di Seppia* (*Cuttlefish bones)*, considering his work as light and dry as a floating cuttlefish bone that, sooner or later, will wash on the beach. Pablo Neruda (1904-1973), an expert shell collector, exalted the shape and colours of the shells in his poems, while Robert Louis Stevenson (1850-1894) became an avid collector of shells during his long stay in Samoa, and wrote: *it is perhaps more fortunate destiny to have a taste for collecting shells, than to be born a millionaire!* It seems evident that a passion for the shells has often strayed into almost maniacal behaviour, as evidenced by stories of collecting (**Chapter 9**). For a particularly poetic literary meeting with a shell we must visit Italo Calvino (1923-1985) in his *Cosmicomiche* (1965). Fascinated and amazed with the great cosmological and evolutionary processes of Nature, the author, in his *The Spiral* tale, imagined the shell as an expression of the desire for individuality of a living being, a desire for recognition and identity that would favour the possibility of being chosen by love. A small role, but molluscs played a part in the history of music too: the French romantic composer Georges Bizet (1838-1875) dedicated an opera to *Les pêcheurs de perles*, while the most famous Irish ballad, *Molly Malone*, tells the sad story of a prosperous saleswoman of cockles and mussels, very popular in Dublin:

In Dublin's fair city,
Where the girls are so pretty,
I first set my eyes on sweet Molly Malone,
As she wheeled her wheel-barrow,
Through streets broad and narrow,
Crying, " Cockles and mussels, alive, alive, oh!"
" Alive, alive, oh,
Alive, alive, oh,"
Crying " Cockles and mussels, alive, alive, oh".

She was a fishmonger,
But sure 'twas no wonder,
For so were her father and mother before,

And they wheeled their barrows,
Through the streets broad and narrow,
Crying, " Cockles and mussels, alive, alive, oh!"
She died of a fever,
And no one could save her,
And that was the end of sweet Molly Malone.
But her ghost wheels her barrow,
Through streets broad and narrow,
Crying, " Cockles and mussels, alive, alive, oh!"

Ringo Starr of the Beatles (1969) wrote the song *Octopus's Garden*, inspired by the behaviour of the animal for collecting stones and shiny objects in front of its hole to create a rocky garden, while in a traditional Neapolitan song, a limpet is one of the main protagonists of a famous "tarantella*", Lo guarracino che jeva pe' mare* (*The wandering damselfish*) (Groeben and Gambi, 1992). He informed the tuna fish that his girlfriend took a liking to the wandering damselfish: the result will be a spectacular underwater brawl, involving, in an extraordinary musical "crescendo", all the marine inhabitants of the Gulf of Naples.

Molluscs were sometimes protagonists even at cinema. Just remember the fight between Captain Nemo and the giant squid in *20,000 Leagues under the Sea* by Walt Disney (1954). Always Nemo, who draws from the sea all the resources he needs, also receives pearls of extraordinary size by giant *Tridacna* clams, in his underwater garden. Pearls were even featured in the film *The Pearls of the Crown* (1937) directed by Sacha Guitry and Christian-Jacque which tells the story of seven pearls that are linked to the vicissitudes of many characters, stirring tales and historical events. In the fairy film *Doctor Dolittle* (1967), directed by Richard Fleischer with Rex Harrison, the eccentric protagonist goes in search of the legendary *Great Pink Sea Snail* and he was able to speak with oysters. And who can forget Ursula Andress, dressed only in a white bikini, collecting Jamaican shells in

Dr No (1962) or the charming Julia Roberts in *Pretty Woman* (1990), dealing with those "bitch" snails at the restaurant!

Octopuses and squids were very often monsters, as in the *Wild Wind* (1942) by Cecil B. de Mille with John Wayne and Ray Millard, which won an Academy Award for special effects. The protagonist of the underwater scenes is a very bad squid, as well as in the film *Tentacles* (1977) by Oliver Helman, with Henry Fonda and Shelley Winters, an Italian film following the disaster-fashion, common of those years.

Figure 4: The fight between a diver and an octopus in a popular Italian magazine (1920).

In fact, cephalopods, such as octopuses, cuttlefish and squids, have often sparked an ancestral fear, a fear that in part stems from the ancient myths, but also from reality (Fig. **4**). Giant squids, the mythical kraken, that were said to attack ships, they really exist and are known since ancient times. In the cinema, the kraken returns in the film series *Pirates of the Caribbean* (2003, 2006, 2007, 2011) with Johnny Depp and Orlando Bloom, while the ship on which the films revolved is named *Black Pearl*. Even in the *Little Mermaid* (1989) by Walt Disney, the octopus does the bad things, and so in *Spongebob*, a successful TV cartoon (1999) where *Squidward*, a squid colleague and neighbour of the protagonist, is petty and whiny, in other words, very annoying.

Figure 5: Drawing from the satirical magazine Punch (London, 1888).

The octopus, despite being a completely harmless animal, because of its tentacles, has always represented evil: just think of the concept of "octopus", a term used to refer to all the mafia and criminal organizations that have many interests and fingers in every pie. In the 19th century, the same British Empire was called the "very octopus of Nations" by her enemies for her aggressive international politics, while in a drawing from the satirical magazine *Punch* in London (1888), the monopoly is represented as an octopus wrapping the fragile boat of free competition with its arms which, in vain, opposes it (Fig. **5**).

Not all octopuses are bad. In Rapallo (Liguria, Italy) a bronze octopus welcomes tourists, while another octopus is remembered in the Tellaro fishing village, still in Liguria. The legend, mentioned also by the Italian writer Mario Soldati (1906-1999) in *The Octopus and Pirates*, told that on the night of 14 August 1660, during a pirate raid, a giant octopus clung to the ropes of the church bells, giving the alarm and saving the village. In its memory, a traditional *Festival of the Octopus* is held in Tellaro on the second Sunday of August, and a plaque on St George Church recalls the event:

ANNO MDCLV - XIV KAL - AUG
SARACENI
MARE NOSTRUM INFESTANTES
SUNT NOCTU PROFLIGATI
QUOD POLYPUS
AES CIRRIS SUIS SACRUM
PULSABAT

Divination skills were also assigned to an octopus: during the 2010 Football World Championship in South Africa, the octopus *Paul* in Oberhausen Aquarium (Germany) became famous all over the world, being able to predict correctly the match results involving the German team, choosing, before the game, food between two bowls, each bearing the team flags! At his death, the Aquarium dedicated a monument to it.

Among the various legends that must be rejected is that the term "ostracism", in ancient Greece the act of banning a citizen, is derived from the word oyster. In reality, Greeks used a crock (*ostrakon*) on which to write the name of the victim: to write on an oyster would not have been easy, even if the name written on mother-of-pearl would provide assurance that no one could erase the choice!

The contribution that molluscs have provided and continue to give to the experimental biologists and medical science is great. For many decades, large marine gastropod slugs such as *Aplysia californica* and squids were used as lab animals: they seem to have been designed by Nature herself to facilitate research on neurology. In addition, nudibranchs, a group of gastropods which has developed a unique defence system based primarily on the production or acquisition of toxic substances, represent one of the main sources to identify new active substances of biological interest (Avila, 1995; Costa Neto, 2006), whose properties were already known by some ethnic cultures. In addition, thanks to some filter feeders such as mussels, we can monitor water quality and quantify the concentration of many toxic elements present in the environment.

Molluscs have had a role in the development of the mathematical sciences too: Archimedes and Descartes, in describing their spirals, were certainly inspired by the plano-spiral shell of the *Nautilus,* wrapping along a plane. It constitutes a perfect logarithmic spiral. From the 16th century onwards, it will be the model of the "golden section" that has been, for centuries, one of the Western canons of the idea of beauty, harmony, and aesthetic perfection. The great American architect Frank Lloyd Wright (1867-1959), who was inspired by just this shell to design the Guggenheim Museum in New York, while declaring himself atheist, said: *I do not understand those who see the presence of God in Nature, I only spell it in the shells* (Sala and Cappellato, 2003).

Moreover, *Nautilus* is characterized by a shell divided into chambers connected by a siphon: varying, by osmosis, the pressure of fluids (gas and liquids) inside them, the mollusc is amending its buoyancy and able to move up and down, without effort, along the water column for hundreds of meters. This complex system of immersion and submersion has somehow inspired the builders of submarines: it is not a coincidence that three submarines, among the most famous in the world, were all named Nautilus: that built in Paris by the American engineer Robert Fulton (1765-1815) in 1798, the literary one of Captain Nemo (1870), and the first atomic one, which first crossed the Arctic Ocean, passing under the North Pole in 1958.

Since the Renaissance, all the shells were very popular, and the subject of collection: they have been and still are the object of desire of many and now the hobby of collecting shells involving tens of thousands of fans around the world. The fashion started in the 17th century in the UK, Germany, and Italy. Afterwards, these private collections will become the main *corpus* of the natural history museums in Europe. Even the fossil shells are very popular and searching is often organized in appropriate areas. One of the most popular in Europe is the West Dorset coastline, between Lyme Regis (Fig. **6**) and Charmouth, in England, an area so rich in fossils to be called "Jurassic coast". A very prosperous market of shells and fossils exists: just browse internet and enter the sites where specialized auctions are organized, to realize the interest they arouse. The great popularity of shells makes them likely to appear frequently on the stamps of many countries, especially in tropics.

The site *www.conchology.be/en/shelltopics/stampsgallery.php* collects images of more than 5,000 stamps on shells, printed in approximately 300 countries!

Moreover, they have also entered into vernacular speech: in Italian, if you define a girl as a "mussel" is not to pay her a compliment, and yet, it is said: "closed like an oyster" or "slow as a snail." And today, the snail-mail now that we have e(lectronic)-mail, has immediate symbolism. The popularity of molluscs has no limits: it is worth mentioning that there are Snail Race Championships organized in Britain and France.

Many stories, more or less important, have involved Man and molluscs: this book will tell how and sometimes when, often with only a small amount of calcium carbonate, they were able, not only to produce an incredible variety of forms and structures, but have played and continue to play an important role in our cultural evolution.

Figure 6: Lyme Regis (Dorset), a street-lighting in ammonite shapes. These fossils have made the place famous.

Figure 7: An African mask decorated with cowries.

CHAPTER 2

Magic, Mythology, and Religion

Abstract: In all cultures, people wore and used shells for decoration, or religious purposes. The oldest record dates back to about 500,000 years ago, and since Palaeolithic, no human grave was found without ornaments made with shells. Over time, some shells even assumed magical properties, often linked to the mysteries of conception, birth and femininity: a symbolism summarized in the myth of Aphrodite, the Goddess of Love.

The tradition of putting one or more shells in the mouth or on the eyes of the dead has been well-known in China as well as in Europe, later adopted by the Greek and Roman cultures. This tradition, so surprisingly common in many traditions, recall the two fundamental attributes that Man has given to some shells: religious symbol and economic value.

One shell above all, the scallop, the shell of St James, has assumed an absolute importance, intertwined with the history of Christianity: it is the symbol and witness of the pilgrimage to Santiago de Compostela. This shell often accompanied the Christian in his grave, according to a millenarian tradition. Other shells, mainly pearl oysters, were the witness of the pilgrimage of Muslims to Mecca, while various religious beliefs, in some way, involved shellfish: in Sicily, with the tradition of, as well as in England, with the legend of.

Finally, some molluscs can produce the pearl, an extraordinary symbol of perfection, beauty, and vitality. According to an old tradition, in fact, the pearl has the power to preserve beauty and sexual desire.

Key words: Aphrodite, Arene Candide Cave, Balzi Rossi Caves, *bestiaria*, Blombos Cave, Concha de Santiago, cosmic prayer, cowries, Cro-Magnon Man, Lascaux Caves, *Dentalium*, Java Man, Los Aviones Cave, molluscs, mythology, native Americans, paper nautilus, *Pecten*, pilgrim's scallop, Pope Benedict XVI, St Lucy, St Hilda, Santiago de Compostela, *Spondylus*, Üçagizli Cave.

Riccardo Cattaneo-Vietti, Egidio Trainito and Mauro Doneddu

Again, the Kingdom of Heaven is like a merchant looking for fine pearls. When he found one of great value, he went away and sold everything he had and bought it.

Matthew, **13**: 45-46

During his complex cultural evolution, Man attributed values and symbolism to some shells, firstly in tribal rituals and subsequently in many religions. The shell, in fact, born from the water, has often symbolized the miracle of life itself or represented the God of water, rain, and wind. In the rituals, they have been used also to decorate masks and fetish-helmets (Fig. **7**), magical objects whose role was *to make visible what cannot be visible*: often they become objects for superstition or real amulets, with strong apotropaic meanings, to ward off evil influences. Superstition still survives today: many people wear a band with some "good luck" cowries: certainly a tribute to a tradition, but also the sign of a deep animistic vision of Nature, still widespread in the world and whose origin is lost in the mists of time (Fig. **8**). Today, in many Pacific Ocean cultures, for example, they still hang shells on the door, an amulet against bad spirits. Of course, there are different species for different locations, but always shells: *Lambis lambis* in the Ryūkyū Archipelago (Fig. **9**), *Nautilus pompilius* in Polynesia, or disks from a great snail, *Bertia brookei,* in Borneo. The latter has become very rare today, probably for this reason. For centuries, sailors have often considered it a sign of good luck to meet the paper nautilus (*Argonauta argo*) floating on the sea surface: one of the first written evidence of this belief can be found in *Halieutica*, the poem on fishing written by the Greek poet and naturalist Oppiano of Anazarbo, who lived in Cilicia under Marcus Aurelius in 180 AD: *Oh precious nautilus, traveller of the sea! Because of you, the man will foresee warm winds. Please, send us a clear sky.*

Figure 8: Even today in Cameroon, the patient brings a padlock, embellished with cowries, to the shaman: he buries it, after having closed the lock and thrown away the key, a symbolic gesture “to close" the disease.

Even in modern times, according to people and localities, snails have positive or negative powers: in Rome, for example, snails are traditionally consumed because their horns are symbols of trouble, and when you eat one of them, the disputes disappear.

Shells in the Prehistory

In all cultures, people wore jewellery for decoration, or for religious purposes. The earliest record dates back to about 500,000 years ago when a Java man (*Homo erectus*) engraved geometric lines on a shell of a freshwater bivalve: the first artistic behaviour by a hominin never observed before (Joordens *et al*., 2015).

Figure 9: Hanging shells of *Lambis lambis* in the Ryūkyū Archipelago, amulets against bad spirits.

After, men, women, and children all wore jewellery of some type. Jewellery was made from bones, animal parts, stone, beads, metal, and obviously from shell. Early in the Palaeolithic Age, shells, especially nassarids easy available inshore, were used as a personal adornment or for magical purposes. It is often difficult to distinguish between the ornamental and ritual roles, but it is certain that many shells in the Palaeolithic have taken on such a symbolic importance as to accompany the dead even in to the grave. In Europe, as well as in America, Africa, Asia, and Oceania, the buried dead are still found adorned with headdresses and loincloths made of thousands of shells, certainly symbols of high social status. Their daily use is, therefore, very old and imbedded in culture for millennia: starting from 20,000 years BC, there is no human settlement where one cannot find ornaments made with shells. Man-made necklaces made with *Nassarius* shells are present in Palaeolithic settlements, dating to 100,000 -75,000 years ago in Blombos Cave, approximately 200 km from Cape Town (South Africa), along the beaches of Indian Ocean, and considered one of the first pieces of evidence of a cultural behaviour of *Homo sapiens*. In the same culture, abalone shells, as *Haliotis midae,* were frequently used as pots for ochre and other pigments (Henshilwood *et al.*, 2009; Henshilwood *et al.,* 2011). Even in Europe, Neanderthal Man, 50,000 years ago, knew how to use *Glycymeris*, *Acanthocardia,* and *Spondylus* shells to obtain pigment containers or for ornamental purposes, as demonstrated by the findings in the "Cuevas de los Aviones" and "Antón" in Murcia (Spain) (Zilhao *et al.*, 2010). In the same way, people that lived 40-45,000 years ago, in the "Cave of Üçagizli" along the Mediterranean coast of Turkey, were able to make necklaces using *Nassarius gibbosulus* shells (Stiner *et al.*, 2002). Many jewellery shells have been found also in Cro-Magnon Man graves, more than 20,000 years ago in the Perigord (France), area famous for the Caves of Lascaux, while in the Gorges de l'Ardeche, La Madeleine, a young man wearing a leather dress, embellished with a thousand fragments of *Dentalium*, was buried nearly 12,000 years ago. In Italy, starting from 30,000 years BC, many testimonials on the use of shells for ornamental purposes as well as rituals were found. In the "Arene Candide", a large cave at Finale Ligure (Italy), a burial among many, known as *The Young Prince* for the richness of his funeral ornaments, was discovered (Bernardini, 1977). The head of this young man, whose death dates between 20,000 and 18,000 years ago, is still adorned by hundreds of pierced shells of a nassarid, *Cyclope neritea*. The same happened in the relatively close "Balzi Rossi" (western Liguria), caves inhabited from the Lower Palaeolithic and where several tombs, dated between 35,000 and 10,000 years old, were found. They are often accompanied by numerous shells (*Patella ferruginea* (Fig. **102**), *Aporrhais, Monodonta, Cerithium, Hexaplex trunculus, Cypraea, Glycymeris*), mostly marine species probably used as food but also for ornamentation. All burials from the Neolithic (Fig. **10**) to the Middle Ages show shells: fragments of shells, coral pendants, and other stones are still present in archaeological sites dating back to 4,300-3,500 BC in Ubaid, Mesopotamia (Sharabati, 1981) or in the North Caucasus (1,500 BC) and in Ukraine. Large quantities of shells were also discovered in Egyptian and Assyrian-Babylonian tombs as votive offerings and were more likely to ensure the welfare of the deceased in the afterlife. In the Americas, the use of shells as decorations in funerals continued to be very widespread also in more recent age: it is extraordinary the *Indian Princess*, a female skeleton dating back to 1,000-1,300 years ago found in a Indian village at Aztalan, Wisconsin, whose dress is still adorned with over two thousand clam shells. The presence of shells inside a burial induced mistakes in archaeologists: *Io spinosa,* a graceful tuberculate fusiform shell from the southern central US streams, found in Indian graves, erroneously suggested that the tribes must have once lived near the shore, but no such shell occurs in the sea. The tradition will continue into Medieval Europe: the scallop (*Pecten maximus, P. jacobaeus*), the symbol and witness of the pilgrimage to Santiago de Compostela, where the tomb of St. James the Great lies, often will accompany the Christian in his grave.

The tradition of giving a value to a shell is very ancient (**Chapter 6**) and well known. To put one or more shells (cowries, in general) in the mouth or on the eyes of the dead has been well-known in China as well as in Europe. This was later adopted by the Greek and Roman cultures: it should be a shell or a coin, but it was necessary to offer a fee to Charon, who brought the souls of the dead across the

River Styx (or Acheron) to Hades, to pay for his trouble. Inevitably this tradition, so surprisingly common in many cultures, recall the two fundamental attributes that Man has given to some shells: religious symbol and economic value.

Figure 10: *Dentalium* shells necklaces adorned bodies in the necropolis of the “Bonu Ighinu" Culture (5,000 years BC), in the Neolithic village of Cuccuruis Arrius, near Cabras (Sardinia).

Old, unsuspected trades

The more a shell is rare and shiny, the higher is its magical and symbolic value, favouring an incredible trade along the coast and from the coast to the interior, and across continents. In France, approximately 25-30,000 years old, some shells not belonging to the European fauna were found in graves dating back to Cro-

Magnon Man, such as *Cypraecassis rufa*, an Indo-Pacific mollusc. Even in the "Grimaldi Caves" in Liguria, on the border with France, a fragment of this shell was found (Jackson, 1917), testifying to its use and distribution. In Liguria, in the "Arene Candide Cave", numerous shells of *Zierliana oleacea,* a gastropod native to the Indian Ocean, were found: all these discoveries bear witness to the great commercial and cultural exchanges that existed in prehistoric times between West and East. These shells, mainly cowries (*Monetaria monetaria*, *M. annulus* and *Cypraea tigris*), coming from Red Sea and Indian Ocean, reached Western Europe by travelling through Arabia and Anatolia, along the coast of northern Africa and the Danube Valley, passing from hand to hand, generation to generation, witnesses of the complex trade that existed between the peoples of Europe and the Near East. The "Incense Trade Route", from South Arabia to the Mediterranean, served as a channel for trading goods such as spices, precious stones, and ebony as well as pearls and shells. These last were often cowries that, in time, would become important tools linked not only to trade, but also to traditional and tribal customs. The cowries are, in fact, a typical metaphor of the female sex, but not only that the shell *Dentalium*, with a sharp sabre tooth-shape, could be interpreted as a phallic symbol: both were often found in prehistoric sites in the Caucasus, in funerary urns in Lithuania, and in northern German burials. Along the shores of the Pacific American, other species of *Dentalium* were used as money for thousands of years (**Chapter 6**). On these local customs and trade, the Europeans obtained a big advantage, so very fraudulent, selling *Dentalium* shells imported by the millions, from Europe and Africa, to the American Indians. Even the modest-looking *Spondylus gaederopus*, the Mediterranean thorny oyster, travelled throughout the Palaeolithic and Neolithic Europe with great success, spreading from Balkans, mostly along the Danube basin and into the Dniester valley in Russia (Saul, 1974). Curiously, also the concave shape of some American thorny oysters, among many other bivalves at their disposal, affected the Native Americans: the rounded and compact shells of *Spondylus calcifer* and *S. princeps*, especially frequent along the rocky shores of the Pacific, were often found in graves of both eastern Mexico and pre-Columbian Peru (Berrin and Larco Museum, 1997). If the *Spondylus* shell was red, it was processed to make necklaces and jewellery, taking on a special ethnic and economic meaning. At the beginning of the 16th century, when the first Spanish explorers arrived in the current New Mexico, they observed that the ornaments of the Pueblo and Navajo (Fig. **11**), the local populations, were often made with discs and red beads that were exchanged for red coral, but in reality made by the *Spondylus* shell. Immediately, the dishonest and greedy Spaniards realized once again a knavish bargain: until the early 20th century, Europeans continued to import large quantities of poor quality red coral to America to be sold to Indian artists who were very interested in working these "stones" in red. Even the thick and ovoid shell of *Glycymeris gigantea* from the Gulf of California, and other related species, had been used for a thousand years by the Hohokam tribes, who lived in 200-1400 AD between Arizona and Northern Mexico, to achieve bracelets.

Figurines of birds, turtles, frogs, and lizards were carved in the shell valves and these designs can still be found in the modern ornaments of the southwest Indians. These items were bartered with the products of the Great Plains (skins, obsidian, woven fabrics) and Mexico (feathers and copper). In North America, *Dentalium* and *Glycymeris*, even the abalone (*Haliotis*) shells, of obvious southern California origin, “travelled” extensively. They have been found in Alaska and along the main trade routes of the American continent: a trade which lasted until the 19th century, especially from Central America and the Gulf of Mexico to the Cheyenne and Comanche tribes who lived in the Great Plains (Whiteford, 2002).

Figure 11: A Navajo necklace made from *Spondylus* shells, 20th century.

Magic and symbols in shells

Over time, some shells even assumed magic properties often linked to the mysteries of conception, birth and femininity: the shiny surface, the inscrutability of the interior protected by the shell, the smooth whorls and especially the shape of the opening of some shells, metaphors of the female sex, and so evident in cowries, are probably the reasons some gastropods have been considered an omen of fertility in primitive civilizations, also associated with the ideas of eroticism and sexual pleasure. For the Romans, the terrestrial snail, whose duration of mating can last for several hours with a complex series of "amorous preludes", was a symbol of sensuality and eroticism: in fact, Messalina (25-48 AD), the Emperor Claudius' wife famous for her sexual appetites, was often depicted seated on a snail (Pozzoli *et al.,* 1825). But more important is the symbolism linked to the myth of Aphrodite, the Goddess of Love. In fact, one of the most important mythological archetypes around the Mediterranean, known as Cybele, Gaia or Gaea, the Magna Mater for Romans, the creator of the universe, and yet untouched virgin mother of Gods, born from the waves, was often linked to a shell. This report was already widespread throughout the Mediterranean long before the Greek myth of Aphrodite obliged. The Goddess of Love, who's first known depiction dates back some 400 years BC (Cox, 1957) and is kept at the Hermitage Museum in St. Petersburg, was born from the sea foam on a spring day in Kythira Island, in Ionian Sea, and carried on a shell to Paphos in Cyprus, the island dedicated her cult. Even today in the Temple of Jupiter in Baalbek, a Phoenician town, in Lebanon, a lying Aphrodite receives from Eros a symbolic shell, depicted in a bas-relief of the Roman period (2nd century AD): Sandro Botticelli, one thousand three hundred years later, will create the eternal the myth (**Chapter 4**). The shell is, therefore, one of the main attributes of the Goddess of Love not just as a "daughter of the sea" as Aphrodite herself, but because it protects the pearl, the classic symbol of incorruptible beauty. In fact, the combination of goddess-shell is also found in some legends of the New Hebrides in the South Pacific and in Mayan and Aztec cultures (Saul, 1974), signifying perhaps a common origin of the myth or, at least, a large diffusion capacity of stories during ancient times. Even the Aztecs considered the shell a symbol of femininity: "Tecsiztecatl", the god of the Moon, who governs the tides, menstruation of the woman, and then synthesizes the ability to create life. For the Mayas, on the contrary, an empty shell is a symbol of death for the previous generation, according to a cosmological cycle of birth, death, and rebirth, while the spiral gastropod seems to symbolize the journey of the soul after death. Love and death, therefore, are inseparable in shell symbolism. And birth, of course: the cowry, for example, with its rounded back to suggest pregnancy and an opening that, somehow, resembles a vagina, is always regarded as a symbol of fertility and is placed in the corresponding region of the genital organs in many African anthropomorphic statues. Even their scientific name, *Cypraea*, not surprisingly, recalls Cyprus, the island sacred to Aphrodite, femininity itself: even today these shells are still a symbol of fertility in many nations in Africa, India, Australia, and New Guinea. Yesterday, like today, pregnant women in Ryūkyū Archipelago, Japan, using to hold a cowry in their hands, shells called "koiasu-gai" or "shell of easy birth", as a sign of good luck. The same occurs in Africa, where objects adorned with cowries were given as presents to pregnant women of the Kuba people, an ethnic group living in the equatorial forests of Zaire.

The lure of sex and femininity was also maintained in malacological taxonomy: when giving a scientific name to a new species or genus, many malacologists have called bivalves and gastropods names like *Venus, Callista, Meretrix, Vulva,* that, somehow, recall the binomial female beauty-sexuality. Some shells bring good luck and are often used by shamans and clairvoyants to practice divination. In Kerala, southern India, cowries are still used to predict the future and their use is also widespread in "voodoo" ceremonies, rituals in the central and south Americas, following an old tradition introduced by the slaves coming from Dahomey. In Brazil, in the "jogo de búzios" is a Candomblé ritual that helps the believer to be invaded by the "orixás", the deities of this magic cult. This tradition echoes the *Santeria*, the syncretic religion that blends Latin-American Christian beliefs and African rites, where the cowries or other similar shells such as the Caribbean *Cyphoma gibbosum*, are used by a shaman in the "terreiro" rites and divination. During the "diloggùn", a rite of foresight, the "pae or mae de santo", the shaman, in turn, throws 16 shells and reads 12. The interpretation, called "modu", depends on the number and the position of the shells falling with the concave part uppermost (Leo Neto *et al.*, 2010; 2011; 2012; Alves *et al.,* 2012). So he can then answer, and predict, just as the Roman soothsayers read the sacrificial animal entrails.

The myth of the pearl and theology

Mollusc can create a pearl, an extraordinary symbol of perfection, beauty, and vitality. According to an old tradition, in fact, the pearl has the power to preserve beauty and sexual desire: it is very likely that the pearl, which Cleopatra dissolved in wine and drank it in front of Mark Antony, her lover, has had this understanding, subsequently taken up by a number of pharmacopoeias in the Far East. According to Eastern tradition, the pearl is also a model of incorruptibility, such as gold and jade, holders of the secret of immortality and with which it is possible to prepare filters of immortality: a belief that was also present in the traditions of medieval Europe (**Chapter 15**). In the Taoist tradition, however, the pearl represents the feminine side of things, the "Yin", one of the two opposing and complementary forces ("Yin" and "Yang", in fact) of the Universe. In this sense, the Chinese would put peals in their graves because their regenerating power could help the soul of the dead to integrate into the great cosmic rhythm. In many philosophical and religious movements, in fact, a shell or pearl necklace is not only an ornament or an abacus to number prayers (such as the Christian rosary or Islamic "tusbax", whose 99 grains symbolize the praises to Allah), but represents, philosophically, all the "eons", the attributes of the Spirit, emanations of God, while the wire running through it is the soul that keeps them together. To paraphrase, the rosary symbolizes the "Great Chain of Being" that binds all things and that becomes, in the Christian as in Buddhist or Islamic faiths, the instrument through which one recites the most perfect of prayers, the “Cosmic Prayer” (Cardini, 1988).

According to a Greek myth, the pearl is the result of a "moment of distraction" of the oyster that is "surprised" by a ray of sunlight or a lightning bolt when the valves are open, thus giving life to a nymph, "perfect creature" by Zeus. If for the Greeks the pearl is born thanks to a sun ray, in the medieval *bestiaria* is told that the oyster opens itself, absorbing the heavenly dew and the sun, moon and star lights, generating the pearl. This myth was reinterpreted in the Middle Ages in a Christian way, when it became necessary to revise the pagan beliefs of the entire Mediterranean basin to include them in a more appropriate Christian context: a huge number of Mother Goodness will thus unite in the figure of Mary, the Mother of God. The pearl, as a representation of Mary, is said to have originated in the *De Margarita* (pearl, in Latin), a treaty on the Virginity of Mary by the Church Father Ephrem the Syrian (ca. 306-373 AD). In the new Christian interpretation, Mary receives, as announced by the Archangel Gabriel, the heavenly dew, or the Holy Spirit, and, in this sense, the pearl and the pearl oyster that encloses and protects it, become attributes of Mary, her most succinct representation. After the Marian tradition, a Christological interpretation was also suggested: Matthew the Evangelist compares the pearl to Christ, the mediator between God and Man. This relationship is told in the *Physiologus*, one of the most well-known *bestiaria* of the Alexandrian period, probably written between the 2nd and 3rd century AD. Herein, the story tells of how the fishermen find pearls using a stone, the agate, a stone which has the property of being attracted by the pearl itself. In Christian interpretation of this myth, the agate is John the Baptist, showing us the spiritual pearl, or Christ. The sea represents the whole world and the pearl-fishermen, the prophets. The two shell valves, the Old and New Testaments, and the light or dew that penetrates the shell, the Holy Spirit (Cardini, 1988).

The Middle Age bestiaria

The medieval bestiaria *are old collections of natural stories, fables, and myths in which, through the description of the animal behaviour, moral examples to be imitated or avoided are proposed: a repertoire for the social life.*
Following the Greek tradition of Aesop in the 7th-6th *century BC and the Alexandrian stories (approximately the* 4th-3rd*centuries BC), the medieval* bestiaria *collected many myths and fantastic legends that, in turn, became the main sources for Charles Perrault (1628-1703), Jean de La Fontaine (1621-1698), and Hans Christian Andersen (1805-1875). Some protagonists of these stories are molluscs with their behavioural characteristics. Leonardo da Vinci (1452-1519) referred to this tradition when he wrote the medieval-style morality story, L'ostriga, il ratto e la gatta (The oyster, the rat and the cat) (H. 51v.):*

Sendo l'ostriga insieme colli al[tri] pesci in casa del pescatore
scaricata vicino al mare, priega il ratto che al mare la conduca.
Il ratto, fatto disegno di mangiarla, la fa aprire e mordendola,
questa li serra la testa e sì lo ferma. Viene la gatta e l'uccide.

(Being an oyster, with other fishes, in the fisherman's house, she was discharged near the sea. She implored a rat to lead her to the sea. The rat thought to eat her, so suggested to her to open and he then bit her. She closed her valves around its head, locking. The cat arrived and killed the rat).

The aphorism emphasizes that one that tries to take advantage of a certain situation, apparently favourable but to the detriment of others, will be severely punished. It is clear that animal behaviour representing the vices and virtues of Man has often been used, long before the Walt Disney stories (1901-1966).

Another early Christian symbolism linked to pearl is in the *Song of the Pearl* or *Song of the Soul* present in the *Acts of Thomas*, an early 3rd century apocryphal text: the Son of the King (Christ) came down into Egypt (the world) in search of a pearl (the soul) guarded by a terrible serpent (the devil). The Son of the King is thus forced to wear a "filthy garment" (the body), which will leave only after regaining his soul, and will finally return to the Father’s Kingdom. Christian thought takes over the old Platonic dualism: the imprisoned pearl in the oyster becomes a metaphor of the human condition, in which the immortal soul is a slave of the mortal body (Caprotti, 2009). St. Augustine (354-430 AD) identified the soul with the pearl itself, and in this sense the legend should also read: the only daughter of St. Hilary of Poitiers (ca. 315-367 AD), named Abra, and fanatically religious, begged her father for a long time to give her the "margarita" (the pearl) which would allow her to receive the joys of Paradise. The saint prayed so long as her daughter was taken from tertian fever, and died (Gay, 1988).

If in the early Christian cultures, the shell was a symbol of fertility, in the Christian tradition it also becomes a symbol of baptismal regeneration and purification. The baptismal fonts often have, in fact, a shell shape carved in, and the shell becomes a witness to the journey of the pilgrim in his journey of inner purification. In some works of art is also a symbol of man's grave before the Redemption, while in other representations it evokes the empty tomb of Christ, after the Resurrection.

Finally, it is not a coincidence that on the coat-of-arms of Pope Benedict XVI (Fig. **12**), a shell is present amongst the other religious symbols, summarizing different meanings within the Christian culture. The first, purely theological, recalls St. Augustine who realized the futility of rationally in understanding the infinity of God, observing a youth on the beach, trying to put all the sea water in a sand pit using a shell. The symbolism is obvious: it invites us to know God in humility due to human inadequacies, but following the inexhaustible theological teaching. Secondly, the shell has been a symbol of pilgrimage for centuries: a symbolism that Benedict XVI keeps alive, following in the footsteps of his predecessor, John Paul II, a great pilgrim throughout the world. Finally, the shell is also present in the emblem of the ancient Monastery of Schötten, at Regensburg in Bavaria, Germany, to which Joseph Ratzinger feels spiritually bound.

Figure 12: Coat-of-arms of Pope Benedict XVI (2005). Below is a scallop which stands for pilgrimages. It refers also to a legend about St. Augustine and the Monastery of Regensburg, where Benedict XVI began his career as a priest.

If shells have often taken on positive meanings in Greek and Roman cultures relating to concepts of purity and holiness, in northern Europe the tradition can, on the contrary, symbolize the worst features of human nature such as vanity or emptiness: the shell is, in fact, an empty object, devoid of life, after the mollusc has left it. We have a great example in the famous *Garden of Earthly Delights* by Hieronymus Bosch (1450-1516), where, among the many other symbolisms, a giant mussel is also depicted in which two lovers can be glimpsed: the shell represents adultery, while the man that carries it with fatigue is the unsuspecting cheated husband (Buzzati and Cinotti, 1966).

Shells and religious tradition

Various religious traditions, in some way, involve shellfish. In Sicily, the tradition tells that St Lucy, martyr of Syracuse in the 4th century, during her martyrdom, tore out her beautiful eyes, throwing them into the sea: they became the red calcareous opercula of the gastropod *Bolma rugosa*, which we still call the "St Lucy's eyes" and are widely used in jewellery (Fig. **13**).

Figure 13: *Bolma rugosa* and its red calcareous operculum.

An English early Christian myth is linked to the ammonites, an extinct group of cephalopod molluscs which were *Nautilus*-like, developed in the Devonian period and extinct around the end of the Cretaceous, whose fossils are often abundant in the Mesozoic sedimentary rocks. The legend tells that St Hilda (614-680), Abbess of Whitby turned a plague of snakes into stones to punish a biblical symbol of sin, often associated with Lucifer, the first of the angels, but also the first sinful and rebellious, or fallen angel. The supposedly petrified snakes are the fossil shell of the coiled ammonites and today local artisans carve snakes' heads onto ammonites the so called "snake stones", and sell these "relics" as proof of her miracle (Fig. **14**). In memory of this legend and St Hilda, an ammonite has been scientifically called *Hildoceras*. Snake-stones are present in the coat-of-arms of the town.

Shellfish are linked to some great religious traditions: in Rome, for example, on June 24, St. John Day, snails are consumed by people as the popular story ensured that for every snail eaten on the night of St. John, a misfortune was averted.

Figure 14: An ammonite carved as a "snake stones".

In contrast, the consumption of shellfish, as well as all seafood, is a serious problem for Jews, because Leviticus (**11**: 9-12, 41-43) tells us:

9 *This is what you may eat from all that is in the water: You may eat everything in the water that has fins and scales, whether in the seas or streams.*

10 *And all that have not fins and scales in the seas, and in the rivers, of all that move in the waters, and of any living thing which is in the waters, they shall be an abomination unto you.*

11 They shall be even an abomination unto you; ye shall not eat of their flesh, but ye shall have their carcases in abomination.
12 Whatever in the water does not have fins and scales is abhorrent to you.
.....
41 And every creeping thing that creepeth upon the earth shall be an abomination; it shall not be eaten.
42 Whatsoever goeth upon the belly, and whatsoever goeth upon all four, or whatsoever hath more feet among all creeping things that creep upon the earth, them ye shall not eat; for they are an abomination.
43 Ye shall not make yourselves abominable with any creeping thing that creepeth, neither shall ye make yourselves unclean with them, that ye should be defiled thereby.

This is a clear prohibition against eating molluscs, crustaceans, and other invertebrates: the same fate for the common snails that, crawling on the ground, are related, in some way, to the serpent, the enemy of Man "par excellence". For Christians, it is not so: but a problem was posed at the beginning of the Catholic Reformation, a question that today would make one smile, but at the time, it had great importance: were the snails to be considered meat or fish and could be eaten on Fridays and during Lent, when the Catholic Church is expected to abstain from eating meat? The problem was solved by Pius V (1504-1572), the great Pope of the Battle of Lepanto, who, in full Lent, wanting to taste a dish of snails, pronounced, according to tradition, the solemn words: *Estote pisces in aeternum! (Be fish forever*!).

The pilgrim's scallop

One shell above all, the scallop, has assumed a symbolic and absolute importance, intertwined with the history of Christianity. Around the 11th century, the scallop appeared in religious iconography as a symbol of the pilgrimage to the Holy Land, just as the pearl oyster, often finely worked, was a witness to the pilgrimage of Muslims to Mecca. In a few decades, this custom spread throughout Europe, probably in reflection of what had already happened in Jerusalem: in particular, the scallop became the symbol of the pilgrim who travelled the *Route of Santiago de Compostela*, the long journey to the tomb of the Apostle James the Greater, discovered by Friar Pelayo the Hermit, in 813 AD in the north-western region of Spain. The *Camino de Santiago de Compostela* across Europe was a moment of great historical significance for Christianity in Europe, and still is today: the way that the pilgrim must take to go to Santiago is shown by a modern road sign, a stylized scallop, and the true symbol of the pilgrimage (Fig. **18**). Arriving in Santiago, the pilgrim, after having venerated the Saint's tomb, drives to Cape Finisterre on the Atlantic coast, the most western point of Europe where, according to medieval tradition, "the Earth ended." As a souvenir, he will collect one of the many shells that the sea leaves on the beach of Galicia. Attached to the hat or tied to his staff, *Pecten* becomes gradually the *signum peregrinationis* for excellence, to which healing powers will be attributed, a good luck charm which no one could do without. In time, it will become the universal attribute of the traveller, taking on a broader symbolism to an allegory of the pilgrimage, a journey from darkness to enlightenment, despair to knowledge (Ripa, 1593). The shell also will accompany the pilgrim in his grave, a witness to his faith. However the collection and sale of the scallop will soon become a profitable business, and Santiago people began to offer religious symbols of any kind but especially shells: the main access road to Santiago will be called *Calle de los Concheros* (from "concha", shell in Spanish). Collections from the nearby beaches will no longer be sufficient to cover the demand and shell traffic begins from the Mediterranean (Girod, 2001)

Figure 15: St James and his saddlecloth ornated with scallops.

The monopoly of the production and sale of this devotional mark (Fig. **16**) soon became a privilege granted by the Bishop, and illegal trade was severely punished by excommunication: the sale of shells thus became an increasingly lucrative arrangement that they even included indulgence and the blessing in the price according to the bad actions of the time. In the 12th century Pope Innocent III (1160-1216) gave the sale of the proceeds of this trade to the Canons of St. Peter's in order to avoid too much speculation, a decision later confirmed by Pope Gregory IX (1170-1241) (Bulgarelli and Melli, 2001). In short, selling shells was a good deal. At the same time, the shape of the shell, the universal symbol of pilgrimage, began to decorate churches, palaces, and hotels along the route, from the Pyrenees leading to the tomb of Saint James, to recall the duty of every Christian to make a pilgrimage to the Holy Places.

In the Isla de la Toja, in north-western Spain, it is still possible to admire a chapel whose walls are completely covered by valves of *Pecten* (Fig. **17**), while in Salamanca, the same carved shells adorn the walls of the *Casa de las Conchas*.

In the 12th century, the scallops began to appear in the heraldry of the nobles who had participated in the Crusades, and become emblem of brotherhoods and religious orders, such as the *Ordre de Saint-Michel*, created by Louis XI (1423-1483) in 1469, whose gold collar is formed by small scallops.

They also distinguished the *Orden de Caballeros de Santiago*, the Spanish military and religious order founded in the 12th century in the Kingdom of León to protect the pilgrims along the roads to Compostela, which were then infested by bandits, and to chase away the Arabs, called "Moros", from the Iberian Peninsula.

In the middle of 500 AD, the scallops will become the *Coquille Saint Jacques*, the symbol of other pilgrimage destinations to Mont Saint-Michel and to the Black Mary of Le Puy en Velay, in the Haute-Loire and, with time, to Jerusalem and Rome. In all the Holy Places, people began to sell shells, maybe not real scallops, but still shells.

This symbol is, of course, also adopted by religious painters: Caravaggio (1571-1610), in his *Supper at Emmaus*, represented one of the two pilgrims with a shell (a cockle and not a scallop, to be precise) pinned to his chest, following the model already proposed by Carlo Crivelli (1430?-1495) who, in the 15th century, represented Saint James as a pilgrim, with a cloak and the inevitable shell. Even St. Rocco de la Croix (1348? -1379?), from Montpellier (France), revered by the Catholic Church as the patron of travellers as well as dogs, is often depicted with a scallop pinned on the cloak.

Figure 16: A painted scallop used as devotional mark of the pilgrimage (Toulouse, France, 20th century).

The legend of the "Concha de Santiago"

By the time the scallop acquired increasing importance it was also supported by some legends. In 844 AD, thirty years after the discovery of his tomb, Saint James the Greater appeared riding a white horse during the battle of Clavijo, whose mantle and saddlecloth were composed of scallops (Fig. **15**). According to tradition, James led Ramiro I of Asturias (790-850) to victory against the Moors of the Emir

'Abd al-Rahman II. Since then, James will be called the "Moor-slayer", and the scallops become his most popular attribute.

Figure 17: The wall of the Chapel of the Isla de la Toja (Galicia, Spain) completely covered from top to bottom by scallops.

Another local story told that at a wedding celebrating at a seaside village near Vigo, some riders were playing on the beach by throwing a spear into the air, trying to catch it on the fly, without letting it fall to the ground. Even the groom took part in this game, but his horse suddenly changed direction, plunging into the sea and dragging its rider with it. They disappeared from view. Witnesses swore that they saw only a bright star moving towards a ship that was approaching the shore. It was the ship in which the followers of the Apostle James brought the body of Saint in Galicia. Shortly after, horse and rider reappeared safely and everyone saw that both were covered with scallops. In memory of this miracle, all the pilgrims who, after travelling the Camino de Santiago de Compostela and venerating the tomb of the Saint returned home, have placed this shell on the mantle or tied it to their travel sticks.

Figure 18: A signpost indicating the way to Santiago, at Saint Paul de Vence (Nice).

Figure 19: Fountain of Trevi, Rome: Triton, son of Poseidon, plays a whelk.

CHAPTER 3

The Trumpet Conchs

Abstract: Since archaic periods, shells were used to produce sounds in many cultures. The conch-trumpet is a musical instrument among the most universal, and at the same time, the most durable. Over the centuries, whelks have become widespread in many Mediterranean populations, and also in Indo-Pacific and in almost all of the Americas, even in distant lands over the sea. Of course, there were different shells for different cultures: *Charonia lampas* in Mediterranean, the "chank" *Turbinella pyrum,* considered sacred in India, the horse conch *Pleuroploca gigantea* and *Strombus gigas* in Americas, *Cassis cornuta* in the Indian Ocean, and *Charonia tritonis* and *Syrinx aruanus* in the Pacific Islands.

According to Greek myths, the whelk sounded the end of the Universal Deluge, while in central Europe, the custom of playing it in the wind, during a storm, survived until to the 20th century.

The whelk is not just a musical instrument popular in many cultures and in many historical periods, but has also achieved a prominent place in the arts, as Triton, son of Poseidon, plays a whelk to announce the end of the Deluge.

Shell-rattles, such as "maracas" and "castanets" sounding by shack or percussion, were mainly used in magic rituals by shamans and sorcerers: their sounds accompany ritual dances, marking the beat, or as rattles, to amuse children.

Keywords: Buddhism, *Cassis*, castanets, chank, *Charonia,* Dakshinavarti, deluge, dung-dkar, Bernini, Hinduism, horse conch, maracas, molluscs, *Pleuroploca,* shankha, shell-rattles, *Strombus*, *Syrinx,* Triton, trumpet conchs, *Turbinella*, valampuri.

Riccardo Cattaneo-Vietti, Egidio Trainito and Mauro Doneddu

Stopped the sea fury and laid his trident,
The God of the Ocean mollified the waters,
called the blue Triton, who protrudes out of the eddies
with shoulders encrusted with shells,
and ordered him to blow his whelk,
because at that signal fall within the waves and rivers.
And he took his hollow and twisted whelk,
the shell which extends itself in a spiral, and
if you give it breath in the midst of the sea, fills
with its voice the coast from east to west.

Publio Ovid Nasone (43 BC- 18 AD), *Metamorphosis* **1**: 7

Shake one, and it awakens; then apply its polished lip to your attentive ear, and it remembers its august abode and murmurs as the ocean murmurs there.

Walker Savage Landor (1775-1864), English poet

S'rigth. It's a shell. I seen one like that before. On someone's back wall. A conch he called it. He used to blow it and then his mum would come. It's ever so valuable. ... He laid the conch against his lips, took a deep breath and blew once more. The note boomed again: and then at his firmer pressure, the note, fluking up an octave, became a strident blare more penetrating than before.

William Gerald Golding (1911-1993), *Lord of the Flies*

Since archaic periods, shells were used to produce sounds in many cultures: breaking the apex of large gastropods, in fact, it is possible to achieve a conch-trumpet, a musical instrument among the most universal, and at the same time, the most durable. Over the centuries, whelks have become widespread in many Mediterranean populations, and also in Madagascar, India, Malaysia, and Oceania and in almost all of the Americas, even in distant lands over the sea, such as Mongolia and Tibet. Of course, there were different shells for different cultures: *Charonia lampas* in Mediterranean (Fig. **20**), the "shankha" *Turbinella pyrum* in India, the horse conch *Pleuroploca gigantea* and *Strombus gigas* in America, *Cassis cornuta* in the Indian Ocean, and *Charonia tritonis* and *Syrinx aruanus* in the Pacific Islands. Shell-rattles, such as "maracas" and "castanets" sounding by shack or percussion, were mainly used in magic rituals by shamans and sorcerers. Sometimes, the valves of a sturdy bivalve, fixed near the hinge, were used as "castanets" by the Algonquian people living along the Atlantic coast of North America (Sachs, 1940) as well as in Indo-Pacific, and some regions of Africa: their sound accompanying the dances, marking the beat, or used as rattles to amuse children.

Whelks

According to the myth, the whelk was discovered by Tyrrhenian, the grandson of Hercules and the founder of the Etruscan people who *concha pertusa buccinavit et pagum convocavit. Unde dicitur tuba tyrrenorum melos* (*he blow the pierced shell and gathered the village. For this reason, it is called the Tyrrheni trumpet*) (Cortese *et al*., 2004). In fact, the finding of different *Charonia* shells with a truncated apex in the Neolithic caves of Liguria, Italy (Issel, 1908; Skeates, 1991) testifies to the ancient use of the whelk: similar evidence, but certainly much later, is seen in the frescos of ceremonies of the Minoan civilization (Karali, 1999) or on coins minted in Agrigento, Sicily, before the 4th century BC. At the end of the pre-Christian age, the hole in the shell was made laterally to give rise to a trumpet with mouthpiece crossbar. Whelks were obviously used by fishermen, but this sound also gathered flocks in the Mediterranean area, the Shinto monks in Japan, frightened evil spirits in south-east Asia, and accompanied sacred libations in the Fiji Islands and funerals in Tonga.

But the whelk is also present in cultures far from the sea: for example in the "Puja", a Hindu ritual, in processions of Tibetan Buddhist monks, and in Aztec dances of the Mexican Indians. Certainly an alarm and challenge sound, but also a call to the Gods to secure their favours: for the Hinduism, its "OM" sound evokes the moment of the Creation, the first "breath of life", the creative vibration or "nada" of God, while for the Buddhism, it is the voice of the Buddha himself. On the contrary, according to Greek myths, the whelk sounded the end of the Universal Deluge. Interestingly, the link between shell trumpet and rain seems to be common in

Figure 20: The Sardinian musician Sandro Fresi playing a whelk.

many other Cultures: born in the water, its sound would have the power to activate waters, to regulate the flow of the tides, to attract rains or to stop them: in Central America, the Aztec priests of Quetzalcoatl, the Feathered Serpent, God of Rain and Wind, used a mouthpiece terminal shell-trumpet, called “tecciztli”, “tepuzquiquiztli”, and “quiquiztli” in their ceremonies. Finally, in central Europe, the custom of playing it in the wind, during a storm, survived until to the 20th century (Sachs, 1940). In some cases, trumpet-shells were also adopted by the Christian liturgy: in the Auvergne, southern France or in the Cathedral of Chieri in Piedmont, Italy, during the Holy Week, the sound of a large *Charonia* replaces that of the bells to call the religious ceremonies (Cortese *et al.,* 2004).

The whelk in the arts

The whelk is not just a musical instrument popular in many cultures and in many historical periods, but has also achieved a prominent place in the arts. In the Greek myth, as Ovid (43 BC-18 AD) reminds us in *Metamorphosis*, Triton, son of Poseidon, plays a whelk to announce the end of the Deluge. This combination is essential and the shell will have an important presence in the mythological scenes of the Baroque period. In fact, Triton is always represented with swollen cheeks, while blowing in one or even two whelks as occurs in the *Triton*, considered one of Rome’s most beautiful fountains, and *Neptune and Triton*, both created by Gian Lorenzo Bernini (1598-1680).

Other whelks still appear in the *Fountain* in the courtyard of Palazzo Massimo Rignano in Piazza Aracoeli and at the *Trevi Fountain* (Fig. **19**), the most famous among the fountains of Rome, designed by the Italian architect Nicola Salvi and completed by Pietro Bracci.

The sacred chank: a lucky amulet

And the sounds of conchs and drums and whelks by thousands began to be heard.
Mahabharata of Krishna-Dwaipayana Vyasa, Bk. 3 Pt. 1

The shell of a large gastropod from the Indian Ocean, *Turbinella pyrum*, called *chank* in India and *dung-dkar* in Tibet (Fig. **21**), is considered sacred in the Hindu and Buddhist religions and is the national emblem of the Kingdom of Travancore. According to a Hindu legend, eight centuries before Christ, the "Vedas", the ancient collection of sacred texts of Hinduism were stolen by a chank. Vishnu, the God protector of the universe, took the form of a fish, so he might go down and recover the lost books and destroy the thieving mollusc. Since, Vishnu is often represented with a chank in one hand, and no Hindus begin to pray without having done an act of worship to the sacred shell. The chank is played by both apex or side holes, depending on the locality. It is often embellished with embossed etchings and precious stones, while the mouth can be in silver or gold, and finely worked. The importance of the chank in the Hindu religion and its mythology is such that any single-shell trumpet is considered a character unto itself with its own specific attributes, just like the "Oliphant", the Orlando horn. The Shankha of Vishnu is called "Panchajanya", but each of the legendary heroes of the Mahabharata and Ramayana epics has a "Shankha" with different a name, often poetic and evocative, such as "Devdutta" (divine gift), "Ananta" (endless victory), "Sughosa" (sweet sound), "Manipushpaka" (precious gem), and many more. The chank fisheries of Ceylon and vicinity had great commercial importance and the shells were shipped to Calcutta and distributed throughout the normal commercial channels. The Hindu artists carved and otherwise ornamented chank; they are then suspended as oil vessels for illumination of temples. The Chinese priests keep these curiously ornamented left-handed shells as sacred vessels in the pagodas, and employed them only for special occasions. The oil for anointing the Emperor is kept in one of these vessels awaiting the next coronation. Most of the chanks collected today are used in the manufacture of bracelets, armlets and bangles, an industry that centres at Dacca.

Figure 21: Indian chanks.

A rude saw, operated by feet and hands both, cuts the shell into narrow rings or segments of circles. These are polished, painted, engraved, and inlaid with precious metals and gems. Elaborateness of ornamentation is seen also where cheaper materials, such as tinsel, spangles, and glass beads, are employed. The Hindu women wear these bangles in great numbers on both wrists and ankles. After death, they are buried with their wearers. Smooth chanks are used to put a high polish upon paper and glazed cloth; their weight and smoothness adapt them for such work, in skilled hands. The "button" is cut from the top, and strung as a bead, or "krantah," into necklaces. These were worn by all the Sepoy soldiers in the East India Service (Rogers, 1908).

Figure 22: The lightning whelk, *Busycon contrarium*, from Florida, exported in India to substitute the true chanks.

Today, this shell, still used in religious ceremonies of Hinduism and Buddhism, is the source of a thriving market in India, Tibet, and Bangladesh over the centuries. The constant requests by the faithful, however, have led to overfishing in the waters of Sri Lanka where it was once common, bringing its natural population on the edge of collapse. Today, by law, only shells with a diameter greater than 7 cm can be harvested.

Sometimes the chank shell is naturally spiralled counter-clockwise, where the turns of the shell are wound backwards and the opening of the shell, holding the apex upward, sits to the left. It is rare, but not impossible, in many gastropods. The shells of the left-handed chank, known as "dakshinavarti" or "valampuri", are particularly sought after and revered in India, Thailand, China, and in Sri Lanka, being reputed to bring good luck to their owners. In the 20th century, to accommodate the continuous request for left-hand shells, more and more rare in the Indian waters, the lightning whelk *Busycon contrarium* (Fig. **22**) from western Florida (Saul, 1974) was exported in India, having a naturally left-handed coiled-coil shell, and now in Hindu temples it is possible to observe, in addition to real *chanks*, also this typical American shell, yet another example of globalization.

Figure 23: *Thatcheria mirabilis* is considered one of the most elegant shells in the world.

CHAPTER 4

In art

Abstract: Shells, often small sculptures of extraordinary elegance, are a continuous source of inspiration since 20,000 years ago, when a Cro-Magnon man carved a stone to simulate the turns of a shell. Crossing the Mediterranean *Cardium Culture* to the Aegean culture, artists were mostly inspired by the sea and its inhabitants.
The spiral shape of the gastropods has been adopted by all the towers in the world, starting from the legendary *Tower of Babel*, while several architectonical styles (Manueline Style, Baroque, Rococo, Neo-Baroque, Liberty) used many forms which refer to marine organisms, and in particular to shells. In the Renaissance, the shells of many molluscs (*Nautilus, Argonauta, Pecten, Tridacna*) become models of inspiration. *Nautilus* shell become a key element in jewellery too: the *Nautilus* cups, with gilded silver, gold, and semiprecious stones, will belong forever to the history of art, thanks to the genius of Cellini. At the beginning of 1600, shells were one of the most common subjects, especially in Flemish paintings, and were sometimes protagonists in 19th century romantic paintings.
Even some 20th century architects were inspired by shells, reinterpreting the eternal forms of Nature, according to different philosophies and approaches. We have to remember Leonardo da Vinci, Antoni Gaudí, Le Corbusier, Frank Lloyd Wright and Jørn Utzon. As architectural elements, shells also became important during the 16th century, especially in the implementation of the *nymphaeum*, following the classical tradition of a peaceful place consecrated to nymphs.

Keywords: Aegean art, Aphrodite, *Argonauta,* Art Nouveau, Baroque, Botticelli, Cellini, *Cardium culture, El Caracol,* Church of Saint-Sulpice, Flemish art, Fountain of Trevi, Gaudí, *Hippopus,* Le Corbusier, Leonardo, Manueline style, molluscs, *Nautilus,* Neo-Baroque, nymphaeum, *Pecten,* still life pictures, Renaissance, Tower of Babel, *Tridacna,* Utzon, Wright.

Riccardo Cattaneo-Vietti, Egidio Trainito and Mauro Doneddu

Una donna non con uman volto
Da' Zefiri lascivi spinta a proda
Gir sopra un nicchio; e par che 'l ciel ne goda
Vera la schiuma e vero il mar diresti,
E vero il nicchio e ver soffiar di venti:
La dea negli occhi folgorar vedresti,
E 'l ciel ridergli a torno e gli elementi
L'Ore premer l'arena in bianche vesti,
L'aura incresparle e' crin distesi e lenti:
Non una, non diversa esser lor faccia,
Come pare che a sorelle ben confaccia

Angelo Poliziano (1454-1494), *Le Stanze per la Giostra*

Nature, above all else, was the most inspirational force of the famous architect Frank Lloyd Wright (1867-1959): he advised students *to study nature, love nature, stay close to nature. It will never fail you.* He did not suggest copying nature, but instead allowing it to be an inspiration. In this sense, shells, often small sculptures of extraordinary elegance, are a continuous source of inspiration (Fig. **23**).
In many cases, the fantasy of Nature seems to exceed that of the artist who is amazed by the geometric perfection of their forms, developed in an infinite number of different interpretations. If we exclude plants and flowers, the shells were, perhaps, the most natural source of inspiration for the artist, a way through which he often tried to interpret Nature, to connect to the Creation that surrounded him, according to the classical law of *Natura Artis Magistra.*

The old Age

The wonder of a shell shape and, at the same time, the spirit of emulation and re-interpretation is certainly very old: in the Caves of Lascaux in Périgord, France, inhabited by Cro-Magnon man than 20,000 years ago, we find a perhaps unexpected object, a false mollusc, a small ovoid stone, carved to simulate the turns of a shell (Delluc and Delluc, 1993), one of the first sculptures in the history of Man. In the western Mediterranean, between 6,000 and 4,000 years BC, contrary to what had happened in the East where a fine painted pottery had been established, the so called *Cardium Culture* developed. It was characterized by the production of pottery, called "cardials" because the edge is imprinted with the toothed margins of the valve of a cockle shell, *Cerastoderma edule*, the common "capatonda", a popular food among the people living along the Mediterranean coasts and lagoons (Fig. **24**). Since that time, the shell will not only be an object of decoration, but also become a tool for decorating.

Figure 24: Pottery fragments decorated using the toothed margin of a cockle (Western Mediterranean, 6,000-4,000 BC).

The same happened on the Island of Crete where Aegean art (2000-1400 BC) was mostly inspired by the sea and its inhabitants: octopuses and painted shells adorn

Minoan and Mycenaean potteries, which spread throughout the Aegean to the Cyclades Islands. Examples of this art, which reached a considerable level of expressive, are the alabaster containers in the shape of *Charonia tritonis* and the many ceramics (1,700-1,400 BC), which was often painted with large octopuses, among algae and corals, which seem to envelop the vase with their tentacles. This icon was dear to the Aegean Culture because the octopus took on a very important sacred meaning: according to the Mycenaean Culture, in fact, it accompanied those who died at sea to the underworld those.

Shells and molluscs are a constant source of inspiration for classic art: the twin volutes in the Greek Ionic order, from the 7th century BC, recall the spiral, a geometry that has always fascinated man and who's most common expression is found in nature in gastropods. It will be up to Archimedes, in the 3rd century BC, to define the mathematical contours of the spiral shell (**Chapter 15**). The same interest in molluscs and their forms can also be found in many central and south American cultures, where the ancient Peruvian and Mexican populations often have them in terracotta. Even the early architects were inspired with shells and in particular to the *columella*, the internal spiral of the gastropod shell, winding around an imaginary central axis: spiral staircases were used both the Old and the New World and are still found in beautiful examples in Renaissance and Baroque periods, such as the spectacular spiral staircase of Palazzo Barberini in Rome, designed by Francesco Borromini (1599-1667), one of the leading exponents of the Baroque architecture, or, 250 years after, the ramp built by Giuseppe Momo (1875-1940) for the Vatican Museum in 1932.

The spiral shape has been adopted by all the towers in the world, starting from the legendary *Tower of Babel*. *El Caracol* (which in Spanish means snail) of Chichen Itza in Yucatan an astronomical observatory built by the Maya between 600 and 850 AD, is other splendid example. Even some Arab towers, built at virtually the same time, were inspired by gastropods shell, such as the minarets of Samarra erected on the banks of the Tigris in Iraq around 850 AD, the *Malwiya of the Great Mosque*, a "spiral" 52 m high, and that of the *Abu Dulaf Mosque*. Curiously, the neoclassical style in Europe will continue to design these forms, but never build them, as the *Tower* by Etienne-Louis Boullée (1728-1799), the French theorist of the neoclassical Architecture (Fig. **25**).

Figure 25: The Tower by Etienne-Louis Boullée (1728-1799). A shell of *Turritella*, a common Mediterranean gastropod, inspired the French theorist of the neoclassical Architecture.

In contrast, the Egyptian civilization in the Mediterranean, more linked to a traditional land farming, seems to be more sparing of art forms that refer to shells, although precious metal necklaces of cowries and quartz have been found in several graves of Middle Kingdom (2052-1750 BC) (Andrews, 1981). Later, at different times, many Mediterranean artists will grace villas with their art, before Phoenician and after the Roman periods, along the coasts of north Africa and Italy, with works often inspired by marine organisms, as evidenced by the mosaics of Sousse in Tunisia, those of Pompeii and Herculaneum, and in the Villa of Piazza Armerina (4th century AD) at Enna in Sicily, where they are often represented by octopuses and cuttlefish. At the same time, the radial and symmetrical ribs, certainly inspired by *Pecten* and other bivalves, will spread quickly as decoration elements: their stylized lines began to appear on Greek and Roman funerary urns and to ornate niches of temples and houses, up to and beyond the Romanesque period. It will, however, be with the European Renaissance and then with the Baroque and Rococo, that the "shell" model will assume a prominent place as a decorative element, and reaching its zenith.

Renaissance and Baroque

At the end of the 15th century, the Manueline Style (Fig. **26**) in Portugal began to use many forms which refer to marine organisms and in particular to shells. One example is the famous *Tower of Belém* and the *Mosteiro dos Jeronimos*, constructed in Lisbon between the end of the 1400 and first half of the 1500 by King Manuel I (1469-1521) to celebrate the journey of Vasco da Gama (1469-1524), the great navigator who opened the way to the East Indies and made Portugal rich. But it is only with the Italian Renaissance that whelks, clams, *Nautilus,* and *Argonauta* shells will become models of inspiration, but the most popular shape as in the classical age, remained the scallop. It is a splendid example, the *Fountain of* *the Bees* by Bernini, in Barberini Square, Rome: it is a giant *Pecten,* whose lower valve was to serve as a water trough for the horses (Fig. **26**).

The valve of this bivalve aroused a sense of stylistic perfection and protection: it looks solid and graceful at the same time and in the Renaissance will emerge as the container to protect something valuable.

Figure 26: An example of the Manueline Style in Lisboa, Portugal (left). The Fountain of the Bees (1644) by Gian Lorenzo Bernini, in Rome (right).

The concept the scallop as protection perhaps reaches its highest style expression in the *Sacred Conversation* by Piero della Francesca (1415?-1492), where the apse of the background is a scallop from which hangs, according to current interpretation, an egg, the symbol of life. For others, the egg might be a pearl, a classic symbol of incorruptible beauty and the most concise representation of Mary. During the same period many painters, including Domenico Ghirlandaio and Andrea di Cione, known as Verrocchio (1435-1488), to name the most famous, represent the Mother of God as sitting on a throne under a scallop. It is with the *Birth of Aphrodite* by Alessandro di Mariano Filipepi, otherwise Botticelli (1445? -1510) that the *Pecten* shell, interpreted as a "nacelle" on which the Goddess sails, reached a place of extraordinary importance in the history of art. This masterpiece summarizes the universal idea of feminine beauty and is one of the symbols of the Italian Renaissance and one of the highest artistic expressions of this famous artist. Among the Renaissance paintings in which shells assume a certain importance should not be forgotten the *Pearl Fishermen* by Alessandro Allori, a frescoe in the *Studiolo* of Francesco I de Medici at the Palazzo Vecchio in Florence, where a *Tonna galea* and a valve of *Pecten,* used by fishermen to hold the pearls themselves, are clearly visible. In Rome, you can admire the *Allegory of the discovery of America* (1570-90) by the mannerist Jacopo Zucchi in which, in a triumph for red coral (unlikely in America, considering that red coral only lives in the Mediterranean and in the Pacific), is a rich collection of shells. Even Giuseppe Arcimboldo (1527-1593), well known for creating imaginative heads made entirely of such objects as fruits, vegetables, flowers, fish, and books, when representing *The water*, used shells, pearls, and octopuses, mixed with fish and crustaceans, metaphorically related to the chosen subject.

The shell is not only symbol of protection, but was also used as a metaphor of a "nacelle", according to the ancient myth, taken from Pliny the Elder and later, by the same Botticelli. However, when Pliny told of the "Venus shell", he was most likely referring to the paper nautilus (*Argonauta argo*) shell. In fact, the female of this cephalopod produces a laterally-compressed calcareous egg-case, in which reside her juveniles (Fig. **27**). This very fragile pale "shell" has a double keel fringed by two rows of alternating tubercles. When the sea is perfectly calm, you might see it floating on the surface, carried by the currents and winds.

Figure 27: According to the tradition, the paper nautilus *Argonauta* uses its two front tentacles, with large membranes, as sails.

Figure 28: Fountain of Trevi in Rome: Neptune, God of the Sea, on a great oyster.

A fantasy, utterly unsupported by scientific observations, tells that the paper nautilus uses its two front tentacles, with large membranes, as sails. Raffaello Sanzio also incorporated the concept of shell -"nacelle" in his *The Triumph of Galatea*: a huge shell, pulled by a pair of dolphins, becomes the triumphal chariot of the nymph, while a great oyster, in the baroque *Fountain of Trevi* in Rome, hosts Neptune, God of the Sea (Fig. **28**). The same image will be taken up again and again in the tradition of Neo-Baroque Style, when dealing with marine issues. The concept of shell -"nacelle" also fascinated Herman Melville (1819-1891): *like noiseless nautilus shells, their light prows sped through the sea; but only slowly they neared the foe* ... wrote the great American novelist to describe the boats of whalers launched in pursuit of Moby Dick. Also in this case, Melville referred not to the *Nautilus* shell, but to an *Argonauta* egg-case (Fig. **27**).

Returning to art, the *Nautilus* shell become a key element in jewellery too: the *Nautilus cups* with gilded silver, gold, and semiprecious stones, will belong forever to the history of art, thanks to the genius of Benvenuto Cellini (1500-1571), the famous Florentine goldsmith and sculptor. Cellini masterpieces will be favoured particularly in Germany and Scandinavia. At the same time, in northern Europe, refined craftsmanship created, fantastic cups using shells of this large cephalopod. Other shells, such as those of the big gastropod *Turbo* imported from East, were used for important vases, cups, and mugs. These shells, mounted in silver and studded with gems for royal drinking cups, are now part of the collections of numerous museums and royal collections.

When large tropical bivalves began to arrive in Europe, a process of synthesis came slowly, but surely: the shape of the European *Pecten* and those of the exotic *Tridacna* and *Hippopus* blend together in the artist's imagination, to form a model which, from the Baroque onwards, will adorn most of the buildings: any visit to a baroque church will illustrate this. *Tridacna gigas* shell from the Pacific Ocean, reaching a length of nearly 1.40 m and a weight of more than 250 kg, had an immediate success as a model of stoups or fonts for holy water, although often the shell itself was used.

In Paris, the Church of Saint-Sulpice hosts two fonts of high Baroque tradition, perhaps the most famous in the world, made by the French sculptor Jean-Baptiste Pigalle. The artist used two huge *Tridacna* valves, gifted to Francis I, King of France, from the Republic of Venice, creating a baroque base full of references to marine forms (Fig. **29**). In 1793, in the middle of the French Revolution, the Church of Saint-Sulpice was deconsecrated and declared a *Temple of Philosophy*. It was feared that the fonts, due to the symbol that represented, would be destroyed. The Revolutionary Commission of the Arts in Paris, to which the famous malacologist Jean-Baptiste Lamarck was part, affixed in the church a sign reading: *Respect for the national ownership*. This act was perhaps one of the first to be taken to protect a public property (Isolani and Manachini, 1996).

The use of this shell as font was so widespread that in France *Tridacna* is called "bénitier" (font). Also in many other churches and cathedrals of Europe, such as Notre Dame of Paris, the Church of Sainte-Eulalie in Montpellier, the Church of Our Lady of Lourdes in Rijeka, Croatia, and the Basilica of San Carlo in Milan, is still possible to admire similar fonts, even if made with shell valves smaller than those of Saint-Sulpice.

Figure 29: One of the two fonts in the Church of Saint-Sulpice in Paris made by the French sculptor Jean-Baptiste Pigalle (1714-1785), using two huge *Tridacna* valves.

Still life and shells

Still life pictures are of commonplace objects, either natural (flowers, dead animals, plants, shells) or man-made (glasses, books, vases, and so on) which originated in the Ancient Greco-Roman and Middle Ages, and became very popular in the 17th century, especially in Flemish paintings. Shells were one of the most common subjects as at the beginning of 1600, as the passion for collecting them exploded across Europe (**Chapter 9**), and many artists devoted their art to this style. Flemish art was famous for its paintings of daily life, as opposed to the sacred and mythological subjects so popular in that time, and strongly commissioned by the Catholic Church as instruments of religious education.

In still life paintings, several distinguished painters, such as Balthasar van der Ast (1593-1657) a pioneer and master of this style, became famous especially in Netherlands. In fact, rich merchants, linked to the East India Company, loving to collect exotic shells, often wanting to have their portraits painted with their "rare pieces". One of the most famous portraits is that of Jan Govertsen in *The Peace and the Arts* by Corneliszoon Cornelis van Haarlem (1562-1638): the rich bourgeois is depicted among his most valuable shells and figures representing the Arts and Sciences. Also the *Cabinet d'amateur* by François Francken II (1581-1642), painted around 1636, of a collection of exotic shells, is well known as well as the *Girl with the Pearl Earring* by Jan Vermeer (1632-1675). In the same period, Rembrandt (1606-1669) etched a *Conus marmoreus*, a work famous in the history of the arts. From a scientific point of view, this etching shows a gross error: the shell is left-handed, showing its opening to the left, as it should be right. This is due to the fact that the printing of the design engraved on copper plate produces a mirror image and inevitably the shell become left-handed. Obviously, Rembrandt was well aware of the problem, so much so that his signature and date are printed correctly. This error is not the first and will not be the last in the history of typography. This famous design will have some impact on the paintings of still life and will be taken over many centuries later, as a tribute, by Giorgio Morandi (1890-1964), an Italian painter, in his *Seashell.* Another famous Flemish Adriaen Cohort (1660? -1707) painted *Six shells on a stone shelf,* by distinguished by the simplicity of his painting, away from the studied and very artful compositions of the period.

Culinary passion for oysters, spreading to Europe between the 17th and the 18th century, and the scent of sensuality emanated by these bivalves also had important consequences in painting: Flemish painters began to represent different tables laden with oysters, as in *Still Life with Parrot* by Davidsz of Jan de Heem (1606-1684) now at the Narodni Gallery in Prague, or *Oysters and glasses* of 1610 (Museo Prado) and *Banquet piece with oysters, fruit and wine* of the 1610/20 (State Gallery of Stuttgart) by Osias Beert the Elder (1570-1624), while young and naughty girls, tasting oysters, will be fixed forever in *Lunch with oysters* by Frans van Mieris (1635-1681) and in *Girl offering oysters* by Jan Steen (1658- 60), both paintings now in the Mauritshuis Museum in The Hague. During the Louis XV age, we could not miss *Le déjeuner aux huîtres* by Jean-François de Troy (Conde Museum, Chateau de Chantilly), a painting from 1735 showing the gastronomic importance oysters had reached during the 18th century. Moreover, in the same period, pottery industry of very fine objects developed, particularly in the North Staffordshire (England). These items, often inspired by Chinese porcelain, had great success in Europe and the most varied shapes of shell were often used as models to create pottery or other objects (Figs. **31, 32, 41**) of great importance.

Arts serving Science

With the invention of movable type printing in the mid-16th century, several volumes of natural history were published, with descriptions of animals and plants. Among the first and most famous, De Aquatilibus *(Fig.* ***30****) was produced in 1553* *by Pierre Belon (1517-1564), a French naturalist, and* Historiae animalium *by Conrad Gessner (1516-1565), a professor of zoology in Zurich: of the latter last, comprising five volumes of more than 4,500 pages was printed numerous times during 1551-58 and again in 1587, after the author's death. Gessner was strongly influenced by classical and medieval mythology and while his work is rich in imaginative iconography, it is still considered the first treatise of zoology of the Modern Age. Even* Conchyliorum Historia *(1685-1692), with more than 1,000 engravings and* Conchyliorum Bivalvium *(1696), both by the English naturalist and physician Martin Lister (1638-1712) should not be omitted. From the 17th century onwards, according to the rules defined by Karl von Linnaeus (1707-1778), more accurate and truthful descriptive iconography become increasingly important but in many cases the engravings also have considerable artistic importance (Caprotti, 2010). Gradually, scientific papers will be decorated with colour plates of great interest, both scientific and artistic: shell depictions by dedicated craftsmen, sometimes known artists and sometimes the scientists themselves, were often produced with considerable skill. In the 19th century, this activity was impressive both in zoological and botanical books as well as in several reports of scientific expeditions, enriched with an iconography. To achieve good results was not, and still is not, an easy task. Describing an animal or a plant, the artist must comply to very specific rules: the image must be done from life (when possible), the shades, colour, and morphological relationships must be maintained in a realistic visual perspective, light must fall from the same angle and details, especially those relating to the description of the species, must be recognizable at first glance. In most cases, a photograph does not seem to be able to depict or highlight the important characteristics of the subject you want to portray, and the researcher still prefers to rely on drawings and paintings.*

Between the 19th and 20th century, an extraordinary family of German glass artists, halfway between arts and science, were able to create shellfish, especially sea slugs (Kessler and Russell, 1978) and cephalopods, in glass models. Leopold Blaschka (1822-1895) and his son Rudolf (1857-1939) were incredible in the manufacture of glass (Hosterwitz, Germany). They produced approximately 4,400 models of plants and flowers in glass for many scientific institutions, as well as a collection of marine organisms, especially planktonic ones. Many beautiful pieces are still visible in the collections of the Institute of Zoology of Wien, in Dublin, at the Cornell University in Ithaca, and in the Harvard University Museum. The aims were essentially educational, and the results were outstanding and probably unique.

Figure 30: A paper nautilus (*Argonauta*) described by the French naturalist Pierre Belon in his *De Aquatilibus* (1553).

The 19th and 20th centuries

Molluscs were sometimes protagonists in 19th century romantic paintings: *The shell* by Adolphe William Bouguereau (1825-1905) shows great delicacy and charm, a typical example of bourgeois painting tied to the exaltation of tradition and family values, but Édouard Manet (1832-1883), the Father of Impressionism, will paint a *Plate of oysters* in 1862, now at the National Gallery of Art in Washington. However, little by little, the language of western art of the 19th century will move away from classic and traditional models, choosing to represent and interpret life in all its aspects. Only towards the end of the 19th century, and especially with the Art Nouveau otherwise known as *La Belle Époque*, few years after the Franco-Prussian War to the World War I, many artists will return inspiration by the forms of Nature and in particular of the shells as decorative patterns, analysing the structural elements and translating them into dynamic and wavy lines in the Neo-Baroque style. It is still possible to observe a good example of this in the buffet of the *Gare de Lyon* in Paris, which was opened in 1901, where the décor is aquatic forms and shells, in full Liberty style. In the same period, the pottery art produced several objects remermbring marine organisms, and molluscs in particular (Figs. **31, 32**).

In the early 20th century, western artists lost interest in the static forms of Nature: Futurism, one of the most avant-garde artistic movements of the 20th century, will, in fact, represent only vibrations, movements, and dynamics. All changed in Europe and America: the attitude of the Eastern artists (in Japan, for example), was very different, more tied to tradition, keeping a strong bond with naturalistic models even throughout the 20th century. In Europe, with Symbolism, art will retain a faint contact with reality such as in the *Birth of Venus* (1912) by Odilon Redon (1840-1916) and *Venus in the Shell* (1930), a small sculpture by Henri Matisse (1869-1954), but abstract art quickly established the idea of interpreting reality or inventing new realities: the same Matisse painted a "conventional" *Still life with oysters* in 1940, but 13 years later, he painted a "snail" that strays far from a factual representation. The same happened in sculpture that surpassed the visual arts, searching in new forms of plastics. An example is *The Head and Shell* (1933) by Hans Jean Arp (1886-1966), now at the Peggy Guggenheim Museum in Venice, while the steel *Scallop* (2003) by Maggi Hambling (1945-), dedicated to the composer Edward Benjamin Britten, seems much more naturalistic.

However, many European artists of 1900's, while fascinated by Surrealism and Symbolism, often loved to reinterpret the forms of nature and, sometimes revisited the masterpieces of the past. This is the case of Salvador Dalì (1904-1989) and his two different *Madonna de Port Lligat* (1949-1950), a tribute to the *Holy Conver-sation* by Piero della Francesca, where the shell of the apse is no longer a refined *Pecten*, but a much more popular *Cardium*, mysteriously painted upside down, while a whelk is sailing around. Filippo de Pisis (1896-1956) deserves to be re-membered, considered one of the masters of Italian painting and sculpture of the 20th century, who was fascinated by shells and painted dozens of them.

Figure 31: A Royal Copenhagen porcelain snail (20th century).

Other artists used the same shells in their works: Joan Miró (1893-1983) inserted a real shell in his *Sculpture Object*, while in Italy, the painter Emanuele Rambaldi

(1903-1968), in his "enchanted magic realism" that characterizes all his work, often referred to marine forms and marine shells in particular, which he included in his compositions. After World War II, pop art will be inspired more and more by the reality of industrial civilization and objects of common use, although some artists in Italy, such as Enrico Baj (1924-2003) and Valeriano Trubbiani (1937-) with their fabled bestiaries, created works referring also to shells. The world of molluscs, especially of fossils, also inspired Ray Troll (1954-), a contemporary pop artist in North America.

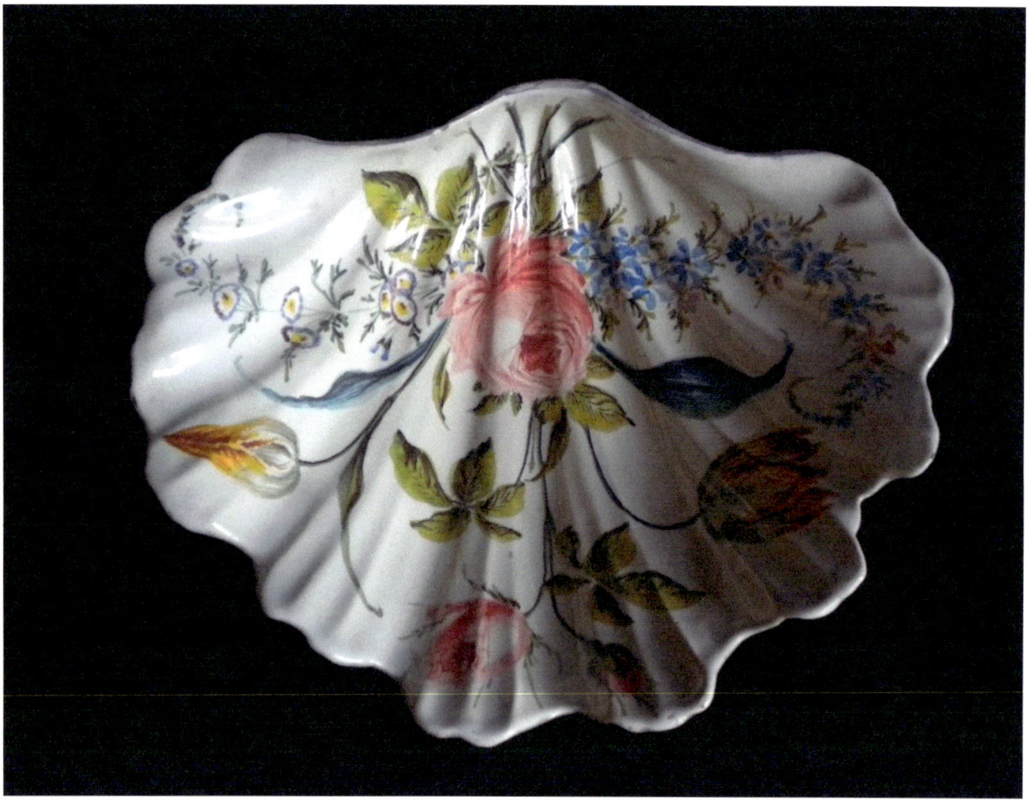

Figure 32: A Bassano (Italy) porcelain bowl (19th century).

In some cases, the use of a mollusc becomes provocative, as in *Djimon with octopus* (1989) by the American photographer Herb Ritts (1952-2002): a sculptural portrait of a black man in which the artist inserts an extraneous element, an octopus, causing a strong contrast between the man's musculature and the octopus's softness, a contrast which, in some way, worries the viewer.

Even some 20th century architects were inspired by shells, reinterpreting the eternal forms of Nature, according to different philosophies and approaches, such as Antoni Gaudí (1852-1926) with his spiral stairs inside the fusiform towers of his *Sagrada Familia* in Barcelona, on which some snails are crawling (Fig. **110**). Le Corbusier (1887-1965) often mentioned the geometric perfection of *Nautilus* in his writings. The idea of an "unlimited growing museum" by Le Corbusier (1939), although never realized, revolutionized the old concept of the 19th century museum and proposed a route in the shape of a square spiral destined to grow over time, in line with the geometrical golden ratio expressed by the *Nautilus* shell. But the most famous representation is the harmonious inner spiral of the main hall of the *Solomon R. Guggenheim Museum* in New York, with six floors leading up to the roof of the building. The Museum, designed by the American Frank Lloyd Wright (1867-1959) between 1943 and 1946 and realized ten years later (1956-1959), remains one of the key symbols of civil architecture of the 20th century and, by admission of the same architect, was inspired by the *Nautilus* or *Thatcheria mirabilis* (Fig. **23**) whorls, considered the most elegant shells in the world. The great architect, when someone asked him why he had used a shell as model, said: *It is the home of a lower order of life, but it has what we lack, an inspired form* (Angeletti and Ferrario, 1983).

Another symbolic building of the 20th century, the *Sydney Opera House* (Fig. **33**), inaugurated by Queen Elizabeth II in 1973, is clearly based on *Lopha cristagalli,* a bivalve common along the Australian reef, or rather on a symphony of valves that open themselves in a harmonious dynamism. The shells-roof projected by the Danish architect Jørn Utzon (1918-2008), which covered the foyer and the concert halls, remains one of the most striking examples of how the geometric purity of the shells may give body to an extraordinary architecture.

Figure 33: The Sydney Opera House is inspirated to the *Lopha cristagalli* shell.

The baroque nymphaeum

Shells became important in art during the 16th century and, at the same time, established themselves as architectural elements of considerable importance, especially in the implementation of the nymphaeum, following a classical tradition of a place consecrated to nymphs, especially those of springs and water. These monuments were originally natural grottoes, traditionally considered habitations of local nymphs. Subsequently, artificial grottoes took the place of natural ones, and were enriched by springs and fountains, sometimes with colonnades, arcades, statues, paintings, and mosaics. During the Baroque period, famous architects designed nymphaeum for *al fresco* summer dining with decorations enriched by three-dimensional compositions that used polychrome pebbles, fake stalactites and, of course, shells (Pizzoni, 1997), such as the *Grotta del Buontalenti*, that the Mannerist architect (1531-1608) realized in Florence on behalf of Francesco I de Medici between 1583 and 1593. Another famous grotto is present in the garden of *Villa del Principe* in Genoa (Italy): designed by Galeazzo Alessi (1512-1572), it has walls fully decorated with shells, corals, ceramic tiles, pebbles, crystals, and fragments of natural stalactites: a multi-material mosaic of exceptional richness. In the 18th century, these nymphaeum became the hallmarks of all European gardens and, later, with natural caves in the 19th century, with a strong romantic inspiration. They are true works of art, such as the style rockery fountain full of exotic shells, built between 1686 and 1695 in the gardens of the *Royal Residence Dutch Het Loo*, near Apeldoorn and the *Amphitheatre of Villa Borromeo* (17th - 18th century) in Lake Maggiore, Italy, dominated by giant stone scallops , like the artificial grottoes of the villa itself. Nymphaeum, grottoes, and fountains thrilled even the French: in Versailles, the ballroom is decorated with a waterfall down the long marble steps decorated with shells, while to the south-west of Versailles, in the *Park of Rambouillet*, the unfortunate Marie Antoinette built the *Pavillon des Coquillages* in 1775, in which not only the walls, but all the furniture were decorated with shells or drawings that reflected their shapes. The use of shells to decorate walls spread to England in the mid 1700's. The extravagant Duchess of Richmond at *Goodwood Park* made a pavilion whose inner walls were completely covered with shells. Not least are the grottos built in Woburn Abbey by the Duke of Bedford or those of St. Giles in Dorset to the Earl of Shaftesbury on the Island of

Guernsey, *Les Vaux Belets*: a small chapel was completely covered, both in and out, by millions of shells, while in Margate, Isle of Thanet at the mouth of the Thames, an extraordinary subterranean complex made up of corridors and rooms completely decorated with shells discovered in 1835. In the Victorian Age, the use of shells to decorate the walls was even more widespread: two small cabins whose walls are decorated with shells are still visible, although not in excellent condition, in Grosvenor Gardens, near Victoria Station in London. Finally, the use of shells in decorating votive shrines diffused in the Mediterranean fishermen villlages since 19th century (Fig. **34**).

Figure 34: The popular use of shells to decorate shrines in Mediterranean fishermen villages (Camogli, Italy).

Figure 35: Shaped-tie collar from the Dani people, Papua New Guinea, 20th century. When the natives met, for the first time, the white men they noted that the commanders wore a tie. Interpreting this as a symbol of power, they began to create their own ties with shells.

CHAPTER 5

Handicrafts and the Kitsch

Abstract: Since Palaeolithic times, the shells of the molluscs, eaten as food, then became small pots or were used to make necklaces, neckbands, bracelets, earrings, or to decorate objects. People used shells available in nature: tusk shells, cockles, scallops in the Mediterranean, abalones, clams, and thorny oysters in Americas, cones and cowries in Africa, pearl oysters and cones in Polynesia. Still today, abalones in New Zealand still adorn many objects with their iridescent nacre.

Many American Indian tribes used mussels, scallops, and clams, queen conch or whelks to create scrapers, knives blades, spoons, while the *columella* of the largest gastropods was used to make earrings. In the Pacific area, the Puka shell jewellery, the use of worked shells to create ornaments, is widespread.

Bivalve valves have been and are still being used as a support for small paintings both in Asia, in America and Europe. Of course, the value of these small paintings depends on the ability of the artist: in Maine (USA) it is still common to decorate the Christmas tree with painted oyster and scallop shells.

The use of shells to achieve the most varied compositions became widespread especially among sailors. At the beginning of the 19th century, especially in Barbados, *Sailor's Valentines* came into fashion: they were wooden tablets, generally octagonal in shape, which show romantic designs and emotional expressions by gluing small shells. Those sailors wore them as gifts from their beloved.

Key words: Abalones, baler shells, *Bursa, Cardium,* clams, cockles,cones, cowries, *Dentalium, Haliotis,* handicrafts, kitsch, *Mercenaria,* molluscs, *Neptunea, Pecten,* Puka jewellery, red whelk, Seba, Sailor's Valentines, scallops, shellcraft, pearl oysters, *Spondylus*, *Turbo*, *Tridacna*, tusk shells, window pane shell.

Riccardo Cattaneo-Vietti, Egidio Trainito and Mauro Doneddu

The men wore on their heads a "lueza", or oval piece of cloth, ornamented with little white shells, which hides the forehead. Most of them had the nasal cartilage perforated, and wore a small cylindrical adornment, made of a white shell or of wood, and called "ztigau".

Luigi Maria D'Albertis (1841-1901), *New Guinea. What I Did and What I Saw.* London (1880)

Among the coastal populations, the use of more varied shells had already spread, as in the Palaeolithic: the shells of molluscs eaten as food then became small pots or were used to make necklaces, neckbands, bracelets, earrings, or to decorate objects. For example, in the collections of the *Pergamon Museum* in Berlin, it is possible to admire a *Tridacna* shell finely sculpted, witness of the high artistic level reached during the Assyrian-Babylonian period. Anthropologists and ethnologists were always amazed and admired many objects made with shells from every part of the world: naturally the more the shell was beautiful and shiny, the more appreciated and promoted as an object of adornment it became (Fig. **35**, **36, 37**). The Dogon in Mali and the Kuba in Congo, the Nagas in India, the Akha in Tibet, the Maya and the Chimu in Central and South America, the Chambri and Abelam in Papua and Melanesia (Cossignani and Poppe, 1994; Cossignani, 2005) are the major ethnic groups who have used and still continue to use colourful and iridescent shells to create artefacts or decorated tools. Each group used shells available in nature: *Dentalium, Cardium, Pecten, Dolium* in the Mediterranean, abalones (*Haliotis*), clams (*Mercenaria*), and *Spondylus* in America, cones and cowries in Africa, pearl oysters and cones in Polynesia and, still today, abalones in New Zealand where their iridescent nacre still adorns many objects. White cones were in great demand among the South Pacific people, especially large ones, to produce armlets. Small white cones are strung into necklaces.

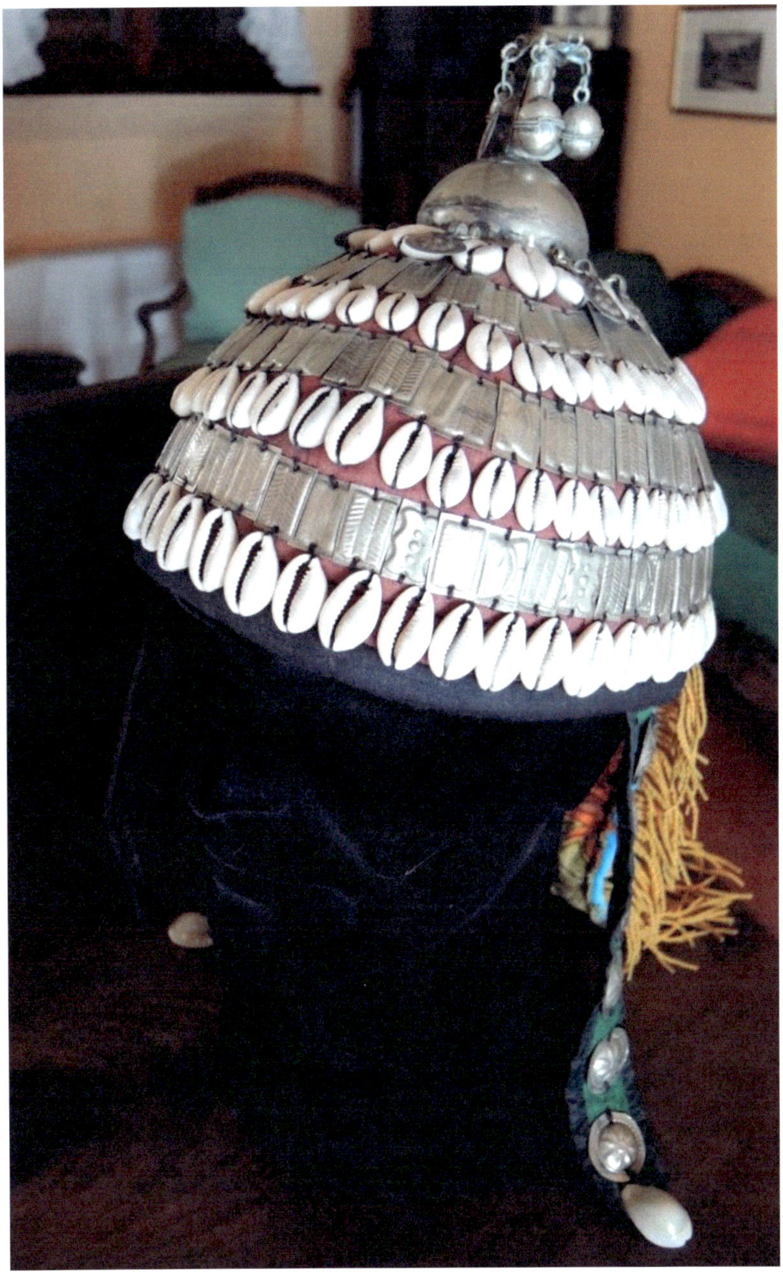

Figure 36: A traditional headdress from Uzbekistan, 20th century.

A great number of mud snails (nassarids) were still used at the end of the 20th century in the Solomon Islands and New Guinea as well as *Chama pacifica* beads to compose wedding belts typical of those populations. Today, many souvenir shops, especially in tropics, sells thousands of *Marginella, Olivella, Littorina, Columbella, Nassa, Cardium,* and *Aequipecten* shells, the smallest of these drilled and strung to produce necklaces and earrings, while others become pots, hanging chandeliers, and curtains (Dias *et al.*, 2011; Gössling *et al.*, 2011). *Monetaria monetaria* shells have always been used in Africa to make necklaces, bracelets, or embellish ceremonial cloaks: the more rich and important was its owner, the more shells decorated the dress. In the Caribbean Sea, *Cyphoma gibbosum* and *C. mcgintyi* replace the Indo-Pacific cowries, for the same purpose. Many American Indian tribes used mussels, scallops, and clams or queen conch (*Strombus gigas*), whelks or *Turbo* to create scrapers, knives blades, spoons, while the columella of the largest gastropods was used to make earrings. According to Bertsch & Marlett (2011), the Seris, a people from Sonora (Mexico), used the opisthobranch *Bulla gouldiana* shells for pendant jewellery, and called this animal *cacaapxon,* or "what fattens (something)".

In India and the Philippines a particular use is made of *Placuna placenta* shell, a widely distributed Indo-Pacific bivalve: its flat, thin and translucent valves, held together by a frame made of wood or bamboo, are used, and lampshades are very commonly sold in Europe. The "window pane shell", this bivalve can produce even small pearls so its natural populations are now over-fished and aquaculture is attempted (Dharmaraj *et al.*, 2004).

Among the Pacific peoples, large *Bursa bubo* and *Turbo marmoratus* were used as oil lamps or cups (today they often become flower pots), until the 19th century. In Europe shells were used to produce an infinity of small object (Fig. **39**): the Shetland Islanders converted the red whelk, *Neptunea antiqua,* up to a length of 20 cm, into "elegant lamps" hanging them in a horizontal position with the lighted end of the wick protruding from the siphonal canal (Rogers, 1908). Baler shells (*Melo* sp.) are used for scooping up liquids, to bale water from small boats. Finally, the giant clams, such as *Tridacna*, were used as bathtubs for children (Fig. **38**), or sold to Europeans for fonts.

Figure 37: A collar from Papua New Guinea, 20th century.

A curious use of large macrids was made in Japan during the Edo period, between the 18th and 19th centuries: pairs of shells, painted with identical miniatures of scenes inspired by classical Japanese literature, were used to play *kai-awase*, a sort of concentration or memory card game, in which players have to find pairs of identical cards. In this case, the equivalent "cards" were decorated with the same drawing (Singer, 1999).

Bivalve valves have been and are still being used as a support for small paintings both in Asia and in America and Europe. Of course, the value of these small paintings depends on the ability of the artist: in Maine (USA), it is still common to decorate the Christmas tree with painted oyster and scallop shells.

Figure 38: A giant clam used as bathtubs for children in the Indo-Pacific area.

Figure 39: Snuff and pill boxes made by shells (England, 19th century).

Towards the kitsch

During the centuries, the use of shells to make small objects produced masterpieces: the extraordinary chisels of Benvenuto Cellini, in the mid-16th century, for example, but was easy to fall into the kitsch, at least according contemporary taste. Already in the middle of the baroque period, some malacological works were enriched with drawings in which the shells were represented forming complex shapes according to the traditional standard due to the Italian Giuseppe Arcimboldo (1526-1593), where natural objects became instruments of a fatuous but aesthetic game (Caprotti 1985). In Europe, the highest expression of this can be found in Albertus Seba (1665-1736), a Dutch chemist and zoologist, who aesthetically influenced all the 18th century. A half-century after his death, a *Thesaurus* was, in fact, published, where he represented his collection of shells with very beautiful and original engravings. Seba's colour plates of shells are arranged in a bizarre style, forming flowers, stars, and other geometric designs, having nothing to do with the rigors of scientific work. However, the book had great success and it is probable that this curious aesthetic style adopted subsequently influenced the taste of the entire 19th century and particularly the Victorian Age. The use of shells to achieve the most varied compositions became widespread especially among sailors. At the beginning of the 19th century, especially in Barbados, *Sailor's Valentines* (Fig. **40**) came into fashion, wooden tablets, generally octagonal in shape, which show romantic designs and emotional expressions by gluing small shells; those sailors wore them as gifts from their beloved. The seafaring tradition of bringing shells back from distant lands was maintained over time and there is no sailor who still does not have these memories of long trips.

Even in Italy, especially in 1800's, shells were used to create floral arrangements, table centre pieces, some of them of great artistic importance. However, this style will lead to mid-20th century and especially in tourist areas a whole series of items of furniture that will tend inexorably to kitsch (Fig. **42**). On this craftsmanship, Louise Travers Allderdice published *The Romance of Shells in Nature and Art* (1962) in which he documents his activities as an artist committed to making paintings, mostly depicting floral motifs, using shells. A few years later, Paula Critchley with *The Art of Shellcraft* (1975) taught others to build the most varied objects using shells: small sculptures, picture frames, or "better" yet, to "beautify" old objects using shells. It is the triumph of kitsch. By-products of cameos, these objects made with the "carcass" of the remaining shells previously used to make the cameo itself.

From the same shell and in particular the big lip you can also obtain balls, barrels, and oval units to create brooches, earrings, and necklaces. Not infrequently they also sell plastic imitations, while plastic religious figurines are placed in large specimens, often illuminated by a light bulb (Fig. **42**).

Figure 40: A painted shell as a Christmas tree decoration, Maine (left). A sailor's Valentine from Cape Cod, Massachusetts, 20th century (right).

Figure 41: An Italian wall hanging copper mold (19th century).

Today the craft, especially if directed to furnishing of houses near the sea, offers an incredible variety of objects in which the shells are "cleverly" used. In many cases, the shells with an important layer of nacre are scraped on the outside to make them shiny and more attractive. The silversmiths also cover real shells with a silver patina, while some houses offer lamps, vases, plates, and other containers decorated with malacological subjects.

Figure 42: Two different kitsch uses of shells: a lamp and a votive composition.

Puka shell jewellery

Puka shell jewellery, the use of worked shells to create ornaments, is widespread in many tropical locations (Fig. **43**). In Hawaiian, "puka" means hole and refers to the natural hole of the shell of a *Conus*, when cut transversely. Puka shell jewellery made from cones has some value, but today most of the puka jewellery is made with other types of shells.

In the Mediterranean, it is still custom to create earrings and other small items, working the "eye of St Lucy", the bright red calcareous operculum of *Bolma rugosa* (Fig. **13**), a gastropod once common in the rocky coasts of the Mediterranean. Even some fossils such as ammonites, often cut and polished to show the internal structure, can be assembled to make jewellery and decorative objects. The use of shells for jewellery and many souvenirs, sometimes with pseudo-artistic aspirations, should always be strictly controlled, especially in tropical environments where the species are threatened. *The Convention on International Trade of Endangered Species* plays a significant control at such borders, preventing the trade of those species that are considered endangered, but ethics and common sense suggest that the fashion of the "natural" and "ethnic" souvenirs should cease in light of a conservationist opinion, strengthened in the public mind at the end of the 20th century.

Figure 43: A "puka" necklace from Sulawesi, Indonesia (20th century).

CHAPTER 6

Shells as Currency

Abstract: Shells have been used for at least four millennia as a bargaining chip in China and Europe, and when metal coins became popular, they were often reproduced on them. The profile of a cowry, a shell always linked to the concept of money, appeared on the first coins of the Western world around the 5th - 4th century BC. The combination of shell and money is therefore very old and many shells, or rather parts thereof, were assembled to form necklaces and belts having a certain monetary value.

They acquired important symbolic and anthropological values, also very much related to social relationships established between various communities.

Cowries are the shells that summarize the concept of money and undoubtedly they played a key role in trades in Africa and in some areas of the Indo-Pacific. This traffic was mainly handled by Arabs, who held it for nearly 700 years. At the end of the 16th century, cowries were an important sign of the wealth in many Western African kingdoms and the basis of the slave trade.

The North America natives knew, for thousands of years, how to work any type of shells, also using them as money: the famous "wampum". Strings of wampum were also used instead of writing, and were created to record important treaties and historical events. Even the ethnic groups from the Pacific coast of North America have their shell currency, adopting the elephant tusk-like shell, stuck in long rosaries, the "haik-wa".

Key words: Coins, cowries, currency, *Dentalium pretiosum,* diwara, haik-wa, kop-kop, kula-ring, money, *Monetaria annulus, Monetaria monetaria,* musanga, *Nassarius camelus,* peak, quahog, rongo, sapi-sapi, tusk shells, wampum.

Better to be paid in cowries. Just so I'm sure I am not fooled!

Gezo (1818-1858), King of Dahomey

Riccardo Cattaneo-Vietti, Egidio Trainito and Mauro Doneddu

Who does not know that some wild, lucky people still use shells for their payments? It is a very comfortable thing; you go at a restaurant and order a dozen oysters, sucks the contents and pay with the shells. The beauty is that each oyster costs a shell; so for ten oysters, you are charged with five shells and the other five you take home, and put aside for retirement.

Achille Campanile (1899-1977), Italian writer and humourist

In the cultural evolution of man, some shells while maintaining deep meanings and magical rituals have also became a common tool of trade (Stearn, 1889). Perhaps it began in prehistoric times, but at a certain point in history, shells became currency. Their use is surprisingly widespread in all continents, and not only among the people who lived along the coast, but also in the inland areas of continents. In China, shell coins was already in use before the 12th century BC in daily trading and it is not a coincidence that the first Chinese bronze coin had a shell shape.

Shells have been used for at least four millennia as a bargaining chip in Europe and when metal coins became popular, they were often reproduced on them. The profile of the cowry, a shell always linked to the concept of money, incidentally appeared on the first coins of the Western world, minted in Lydia (modern Anatolia) around 670 BC, while the drachmas of the 5th - 4th century BC often depicted a mussel, probably *Mytilaster minimum*, perhaps a Magna Grecia symbol (Russo, 1999). Other profiles of shells important for the economy, such as the Mediterranean *Pecten* and *Pinna*, along with that of the octopus, then began to appear on coins minted in Anatolia, in the Sea of Marmara and in eastern Sicily as early as the 3rd century BC (Jacob, 2009), while some coins of the Roman Republic (240-220 BC) minted in bronze relief, show a scallop (Fig. **44**).

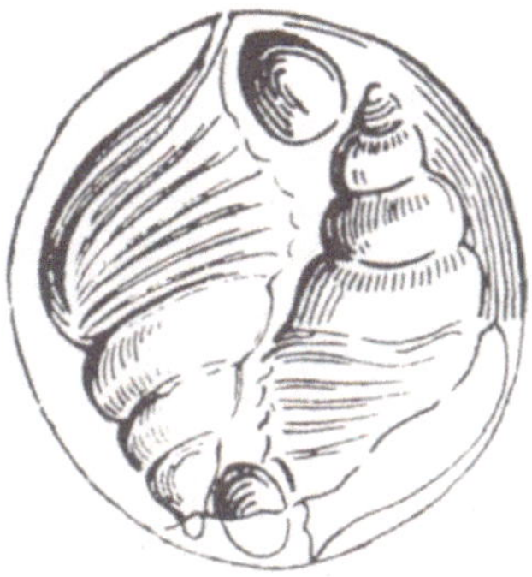

Figure 44: A minoic seal (left), and a Roman coin (right), depicting shells.

Even the shells of some *Murex* shells, from which the purple dye was extracted, appeared on very earliest Phoenician-Punic coins of Tyre in the 5th - 4th centuries BC, and later reappeared again on coins in Imperial Roman times.

Even today, many shells appear on coins minted especially in tropical countries. The "Queen conch" (*Strombus gigas*) is on the back of the silver dollar of the Bahamas, while *Turbinella pyrum*, the shell sacred to the Hindus, appears on the coins of the Himalayan Kingdom of Bhutan. A "spider conch" (*Lambis*) and an octopus appear respectively on the coins of 1 and 50 cents of Tuvalu, a small Polynesian archipelago, while a *Charonia tritonis* adorns the coin of 2 vatu in the Republic of Vanuatu, in the southern Pacific Ocean; the "imperial volute" (*Cymbiola imperialis*) is depicted on the 1 cent coin of the Philippine pesos and an oyster pearl is on a 1,000 pesos banknote. Finally, the 200 Moroccan dirham show *Bolinus brandaris*, the muricid from which once the purple dye was extracted, a flourishing activity in Morocco during Roman times (Fig. **45**).

Figure 45: *Bolinus brandaris*, the muricid from which the purple was extracted, stamped on the 200 Moroccan dirhams.

The combination of shell and money is therefore very old, but not all shells were fit for purpose: the reason for the extraordinary success of some of them as money is easily understood. To become a coin, a shell must have certain characteristics: to be readily available (but not locally), compact and consistent, lightweight and robust at the same time, not altered by continued manipulation. It must also be "attractive" and have an emotional impact that is "positive" towards those who use it: in other words, be glossy or polished. Some shells, as *Dentalium pretiosum*, *Mercenaria mercenaria* (Fig. **46**), several species of *Conus* and especially cowries undoubtedly possess these requirements, and there is no culture in the five continents that have not used one of these in the past.

Figure 46: *Mercenaria mercenaria.* This clam, locally known as quahog, was the basis of the monetary system among the NE American Indians.

In fact, before the invention of metal embossing, it was not easy to find an object that contained all these features in nature. In almost all cultures, shell-money has also undergone processes of cutting, drilling, sanding. Many shells, or rather parts thereof, were also assembled to form necklaces and belts as well as having a certain monetary value, and acquired important symbolic values, very much related to social relationships that were established between the various communities. Necklaces and belts, depending on the ethnic groups, of course, take different names. In Zaire, the Lega people called them *musanga*, while in Malaysia the hard shells are the *sapi-sapi*. In New Britain (Bismarck Archipelago, Papua New

Guinea), the natives, using *Nassarius camelus*, a gastropod common in the mangroves, led their *rongo* or *diwara*, huge necklaces, as thick as 20 cm, to form giant rings up to a meter and a half in diameter (Saul, 1974), the treasure of a whole tribe.

In the "kula ritual", bracelets, and necklaces made of shells were exchanged in a ceremony rich in anthropological meanings between the peoples of Melanesia. In particular the "kula ring" was a symbolic exchange of shell necklaces and bracelets among the communities of 18 islands of Melanesia, east of New Guinea. The basic purpose was to strengthen good relations among people who had sporadic contact as, inevitably, the oceanic trips occurred exclusively with long canoe. Red nassarid necklaces (called *veigun* or *soulava*) were traded with the inhabitants of the northern Islands, while those who lived in the southern Islands received in exchange white shell bracelets, called *mwali,* formed mainly by *Conus millepunctatus*, or sections of *Pinctada margaritifera*. The exchange could take place only between different objects: bracelets, necklaces, and vice-versa. Jewellery would circulate continuously, remaining in the hands of the owner only for a limited period of time, and were traded during visits of the inhabitants of the islands periodically. According to Bronislaw Malinowski (1884-1942), the father of social anthropology and author of *Argonauts of the Western Pacific* (1922), the exchange of these ritual objects, rich in magical meanings, also had the task of building serious relationships between the Islanders, and were the basis for the next trades.

In the 19th century, thanks to *Ovulum angulosum*, a white porcelain cowry from Tonga, valuable cargoes of sandalwood were obtained in some Pacific Islands for the Chinese market and, in the first instance, from Vanuatu Archipelago, 500 km NE of New Caledonia. This shell was not present there, but the natives so highly esteemed it as an ornament that for one shell they would give in exchange a huge amount of sandalwood (Rogers, 1908).

Among the so-called "indigenous coins" unique pieces, such as disks of approximately 12 cm in diameter by cutting clams in the Solomon Islands, or large calcareous opercula of some gastropods, the "cat eyes", used by Papua people, are well known. Among the NE America Indians from, the *wampum* made by beads from the hard shell-clam *Mercenaria mercenaria*, locally known as *quahog*, was the basis of the monetary system, while the Hupa and other tribes of the Pacific coasts utilized the *haik-wa*, a necklace of tusk shells, *Dentalium pretiosum*. Other coins, made using smaller shells or pieces of them, were called *kop-kop*. Among the natives of Central America, a very important role was played also by *Spondylus princeps*, a thorny oyster called *mullu* in Quechua, sought after for the brilliant red colour of its shell. Its use as a currency has been evidenced by the fact that it was part of the Aztec tributes paid to the Emperor Montezuma (Saul, 1974). Maybe even before 2,000 BC, these *Spondylus*, together with *Strombus* shells, formed the basis of intense commercial traffic that had been established between Peru and Central America. The inevitable decline of shell-money, after thousands of years of history, began in the 19th century, with the intensification of trades between East and West, when the Western financial world could not tolerate an economic universe separated from pounds or dollars. It is not surprising, however, that still in the 20th century, among the oldest ethnic groups of Africa, America, and Oceania, living at the edges of the common trade flows, small shells represented the most common raw material, after the metals, to produce coins. Even today in some areas of Papua New Guinea, the shell is informally accepted as an element of exchange; and, the paper currency of the country derives its name from two old shell currencies, *kina* or *keena* (ie, *Pinctada margaritifera*) while its cents are called “toea” or “toya”, such as those once made with cones.

The cowries

Cowries are shells that summarize the concept of money and undoubtedly played a key role in trade in Africa and in some areas of the Indo-Pacific (Polanyi, 1980, 1987; Hogendorn and Johnson, 2003). Cowry, a name of Hindu origins, refers without distinction to two small shells, *Monetaria monetaria* (Fig. **47**) and *M. annulus* (Fig. **48**).

Figure 47: *Monetaria monetaria*, a cowry shell.

They are small egg-shaped shells, colour ranging from white to pale yellow-green, sometimes with thin ochre or orange ring that adorns the back. Both live in the Indo-Pacific and Panamanian regions, where it is easy to collect them in shallow water. Easy to handle, rounded, shiny, and sturdy, they are simple to use and cannot be faked: these are the main characteristics of the cowries and the reason for their extraordinary fortune. The cowries were already in use in many parts of India, especially in Bengal, before Alexander the Great, and were common currency China in 2,000 BC. They were still used in the Middle Ages, as Marco Polo (1254-1324), the great Venetian explorer, witnessed in the 13th century. According to the tradition, he called these shells "porcelline" which means "piggies", because of their shape that resembles a vulva. Consequently, the Chinese pottery, as smooth and shiny as the cowry shells, was called "porcelain". The main areas for shell collecting were East Africa, Maldives, Philippines, and Malaysia.

Figure 48: *Monetaria annulus*. Another cowry shell.

In the history of navigation, the Maldives archipelago has played a key role along the trade route during the monsoon season, in touch the SE Asia and China, the Persian Gulf and E Africa. Maldives were a major resting point for supplies and ship repairs, considerable skills which the Maldivians acquired over time. Around the 9th century, a stop in the Maldives became also a pretext to collect large quantities of cowries, to be used as bargaining chips in the whole of SE Asia, E Africa, and Arabian harbours and, after a long journey, finally, in the West African kingdoms. This traffic was mainly handled by Arabs, who held it for nearly 700 years. Of course, away from the natural habitat of cowries, their use was much more widespread and their value stable and high: among the West African people, in an area where these molluscs are not present in nature, cowries had a huge and constant diffusion: in the 7th century BC, they were already the main currency and reached maximum importance in the Middle Ages.

Arab ships arrived on the coast of the Red Sea laden with cowries, and merchant sailors gave way to trans-Saharan caravans to reach the West Africa markets. And so the peoples of the Sahel, already around 1000 AD, began to use them as bargaining chips: each shell covered an immense distance, such that it was protected from the risk of inflation. At the end of the 16th century, cowries were an important sign of the wealth in the African kingdoms of Mali, Dahomey, and Nigeria, from Timbuktu to Lake Chad, and were the basis of the slave trade: the European slavers paid for slaves bought by African raiders in cowries. The same Emperor of Mossi, the people who lived along the valley of the Volta River, Burkina Faso, claimed by his vassals an annual tribute of one million of cowries.

20-30 million of shells made a rich and powerful man in West Africa. In Uganda you could buy a cow with 2,500 cowries, and an ivory tusk, weighing nearly 28 kg, was worth 1,000 cowries. A goat costs 500 cowries, a chicken 25, and a pipe from 50 to 100 cowries. The possibility of paying for the slaves in cowries led the European slavers to collect large amounts of these shells in the East Indies: in 1848 and 1849, 60 and 300 tons respectively of cowries were transported to Liverpool, to be subsequently sent to Guinea. A criminal increase in monetary supply which caused inflation in a short time, destroying an economy linked to the cowries. A witness to this fraudulent traffic is the wreck of the brig *Glendowra,* sunk in 1873 along the coasts of the British Cumberland, carrying a huge load of cowries collected in the Philippines. For many years to come, large quantities of cowries were found in good condition on the site of the wreck.

Until the First World War, these shells had a monetary spread far wider than the dollar or the pound sterling in Africa, and throughout the 20th century, cowries continued to be used as small change in the markets of Benin, Ghana, Ivory Coast, and Burkina Faso, where local merchants more willingly accepted cowries than national currencies. Until 1921, the cowries remained the official currency of the Maldivian State, when they were replaced by the rufia.

Even today, the “cedis”, the currency of Ghana, is called "cowrie" in the language of the Akan, one of the ethnic groups in West Africa and the profile of a cowrie, *Trona stercoraria*, appears on several coins minted from 1979 to 1996. Even in Mali and Benin, it is easy to find stylized decorations that recall the cowries, especially in bank offices.

The New World shell coins

When the first Europeans arrived on the shores of North America, the natives knew, for thousands of years, how to work any type of shells, also using them as money. On the east coast, in Virginia and New England, they used grains and disks obtained from various bivalves, but especially from *Mercenaria mercenaria*, locally called "quahog". This name is derived from the Native American Narragannset word “poquauhock” which means “horse fish” (Ristine, 2009). The commercial use of this shell was so widespread and popular that when Karl von Linnaeus described it in 1758, he wanted to emphasize its use, hence calling it *Mercenaria* which means hireling. Using shell beads, the Iroquois and the Algon-

quian tribes of the north-east created sophisticated belts with beautiful drawings (Whiteford, 2002) that the settlers of New England called "wampum", while in Virginia they were known as "peak" or "peag" (Verrill, 1939). To produce these belts required much time and their value was reflected the integrity of the shells and the type of shell used. Indeed, there are at least two types of "wampum", depending on the raw material: those obtained from "quahog", whose inner edge is adorned by a purplish blue or dark band, are called "black wampum", and have a higher commercial value. The "white wampum", of lesser value, is instead achieved by working the *columella* of some whelks, such as *Busycon carica*, *B. contrarium,* and *B. canaliculatum,* so common along the coast from the Massachusetts south to Florida. Three beads were worth an English penny in early colonial times, while a fathom string was worth five shillings. The "wampum" has been more a symbol of friendship and mutual trust than a currency, and during the 18th and most of the 19th centuries, it was the main gift exchanged between Indians and Europeans: today, it retains a high cultural value because its designs perhaps tell the story of the Iroquois nation, evidence of a now extinct culture.
An American Indian historian, William James Sidis (1898-1944), wrote in 1935: *the weaving of wampum belts is a sort of writing by means of belts of coloured beads, in which the various designs of beads denoted different ideas according to a definitely accepted system, which could be read by anyone acquainted with wampum language, irrespective of what the spoken language is. Records and treaties are kept in this manner, and individuals could write letters to one another in this way.*

The Pacific haik-wa

Even the ethnic groups from the Pacific coast of North America had their shell currency. In California, the Chumash, skilled fishermen, one of the few Indian tribes who dared to venture into the open sea, called the purple olive *Olivella biplicata anchum,* which has been their main currency for at least 9,000 years. These strings of shell beads were called "kol-kol" (Rogers, 1908). But it is certainly the "Indian money tusk" which took on a role of great importance for the whole North American economy. Yurok, Hupa, and Karok people adopted, in fact, the shells of *Dentalium pretiosum* and *D. hexagonum*, as the currency, scaphopods with a white elephant tusk-like shell, approximately 3 cm long, relatively common in coastal sands of North Pacific and in particular in the waters of Vancouver Island, where the people of Nootka harvested them and traded in California at least through the 19th century.

The *Dentalium* shells were stuck in long rosaries, the "haik-wa", the value of which varied according to the quality of the shells. A rosary of 25 of these shells could be worth nearly $ 250 at the end of the 17th century - it was possible to buy a canoe with a "haik-wa"! This coin was widely accepted along the Pacific coast from Alaska to Baja California, but also in the Great Plains by the Sioux and Comanche, east of the Rocky Mountains. They used it more as ornament than money, creating hats, belts, shoulder bags and other objects. With the arrival of the Europeans, the shells gradually lost value as a bargaining chip, because of large imports of shells in Europe and Asia. The Dutch East India Company and the Hudson's Bay Company, with the same technique used in West Africa, imported large quantities of *Dentalium vulgare* from Europe to New Jersey and *Monetaria monetaria* from the Indian Ocean to Canada. These fraudulent imports caused, once again, a real inflation, defining the end of the shell-currency in the North America.

PEARL-FISHING.

Figure 49: A 19th century view of the pearl fishery.

CHAPTER 7

Pearls and Mother-of-Pearl

Abstract: The pearl, a rare product of some marine and freshwater bivalves and gastropods, has been considered a talisman of fecundity, and become symbol of perfection and absolute purity in many cultures. Greek symbol of love, the pearl also took several religious significances, and remain one of the principal ornaments of the crowns of many royal houses.

The pearl is linked to love, but at the same time, to pain: this is a concept still present in the collective imagination and according to popular belief, pearls lead not only love, but also tears for their owners.

The most famous pearls, but not always those with the greatest value, are due to different species of pearl oysters, living in the Red Sea and Western Indian Ocean as well as in Australia, and Central America. The areas of great tradition for pearl fishing were the Persian Gulf, Red Sea, the Gulf of Mannar in South India, Sri Lanka, Madagascar, and the Malay Peninsula.

With the discovery of the New World, pearls of great beauty were found in the waters of the Gulf of Mexico and along the coast of Venezuela as well as in the Pacific Ocean, in particular, from Australia to French Polynesia and Philippines. In any part of the world the pearl fishing occurred, it has been a story of destruction. The discovery of pearl banks in the tropics provoked, in fact, a kind of "gold rush" in a very short time, with a devastating exploitation not only of the banks themselves, but of the natives themselves.

Keyword: Abalones, *Anodonta,* calcite pearls, cameos, Cleopatra, cultured pearls, fake pearls, freshwater pearls, *Haliotis,* mabe, *Margaritifera, Melo,* Mikimoto, molluscs, mother-of-pear, nacre, pearls, pearl fisheries, Pearly Harvest Festival, Pearly Kings and Queens, *Pinctada, Pseudanodonta,Pteria,Unio.*

Quippe inter scopulos maior pars invenitur, in alto quoque comitantibus marinis canibus, nec tamen aures feminarum arcentur.

Pliny the Elder (23?-79 AD, *Naturalis Historia*).

In sterquilinio, margaritam reperit.

(In the shit, you will find pearls)

Latin proverb

La plus belle perle n'est donc, en définitive, que le brillant sacrophage d'un ver.

Raphael Dubois, French zoologist (1849-1929)

Although the majority can be found among rocks, in high seas and among sharks, you cannot keep them away from the women' ears. So wrote Pliny, speaking about pearls. And in fact, there has never been a natural product more attractive and desired. The pearl has always been considered a talisman of fecundity and become, over time, symbol of perfection and absolute purity, a prerogative of Chinese, Indian, Egyptian, and Babylonian kings and emperors, from Solomon to Cleopatra. For example, the famous mosaics in the Basilica of St Vitale and St Apollinare in Ravenna, Italy, show the Emperor Justinian I of Byzantium (482-565 AD) and Empress Theodora with rich headdresses and pendants adorned with rows of splendid pearls. According to Pliny, among the Roman matrons a "parure" consisting of earrings and rings, each with two or three beads, was in vogue which, colliding with each other, produced a characteristic noise: the so-called *crotalia* or rattles (Battiato, 2012b). In reality, they adorned the Egyptian and oriental women already some thousands of years before Christ: the oldest known pearl was found in Japan in the first Jamon period, approximately 5,000 years BC, while the Louvre Museum preserves a "parure" of pearls from 5th century BC.

Greek symbol of love and marriage, the pearl also become a religious symbol in the Christian world, but will always remain one of the principal ornaments of the crowns of many royal houses, embellishing the neckline and the hairstyles of queens and famous women (Fig. **51**). During the Belle Epoque, pearl hairstyles were all the rage: Mata Hari (1876-1917), the famous dancer and Dutch secret agent, is almost always depicted with complex exotic finery, rich in pearls.

Rare products of some marine and freshwater bivalves, the pearl can also be produced by gastropods and, much more rarely, cephalopods. It is the final result of a reaction of the mollusc to an adventitious presence of a foreign body (not necessarily a grain of sand as it is believed, just the mere presence of a small parasite, as pointed out by the French zoologist Raphael Dubois), which tends to encapsulate it in a lump of calcium carbonate. Sometimes, the presence of a foreign body is not even necessary: the pearl can just as easily be the result of a mantle malfunction and the tissue that produces the shell deposits in a wrong place.

Pearls are generally silvery-white, but there exist in every form: they are often pear-like, while the newer ones are called irregular baroque. Many are lumpy or too small to be worked and not appreciated in jewellery. It is rare but not impossible to find many small pearls in the same oyster, particularly in *Pinctada radiata*: sometimes they grow so close to each other to melt, as has happened in the famous "Great Southern Cross" discovered in 1874, probably in Roeburn, Western Australia: a set of nine pearls fused to form a cross nearly 4 cm long, now part of the treasure of the British Crown. There are two main types of peals: natural and cultured. The first ones are much more rare and precious. Within each category, pearls are considered on the basis of various parameters. The variability in size, colour, and shape is so high that the Romans called the larger pearls *uniones* (*unus* in Latin means unique) as it was practically impossible to find two alike. Their economic value is however not exclusively linked to the size: the colour and brightness are important and depend on the ability of reflection and refraction of light by the various layers of mother-of-pearl. Often the colour is also linked to the location in which the pearl is born: that of the Persian Gulf, for example, differ from those of the Red Sea, called "Egyptian" because they are silver-gray. There are also black pearls, blue, yellow, gold, pink, and iridescent: it is dependent upon the species of mollusc that has formed it and pigments secreted by its mantle. The most famous pearls, but not always those with the greatest value, are due to different species of pearl oysters, belonging to the family Pteriidae: *Pinctada margaritifera* is widely circumtropical, *P. radiata* is from the Red Sea and Western Indian Ocean, *P. maxima* has a distribution extending from the coast of Australia up to and beyond the Philippines, *P. fucata* from the Indian Ocean and the Western Pacific, and *P. mazatlanica,* present mainly along the California coast. Some species of the genus *Pteria*, the so-called "winged oysters", can produce pearls. They are known especially for blister and “mabe” production. In contrast to the pearls, the blisters grow attached to the inside wall of the shell, forming a kind of bubble that is produced by the mollusc to defend itself from a parasite that attempts to pierce its shell from the outside. If they reach considerable size, these growths have some value because can be used to produce earrings or “parure” (Fig. **50**). Depending on the size, an oyster can take six years to form a blister. In technical terms, a blister obtained artificially is called “mabe”, and is produced by introducing a small hemisphere in contact with the inner part of the shell, but in Amami-Okinawa Region (Japan) blisters from *Pteria penguin* are called “mabe” too.

Some Pacific scallops (*Nodipecten nodosus, N. subnodosus*) produce, albeit rarely, non-nacreous pearls which are usually purple. It is possible to find them also in the Mediterranean noble pen shell, *Pinna nobilis*, the largest Mediterranean bivalve.

Figure 50: A parure with "mabes". French School (19th century).

Its shell colour is light brown and the inner surface pearly brown, and it can reach 80-90 cm in length. Widespread in the Mediterranean, from a few meters up to 40 m deep in sand and within the *Posidonia* meadows, was harvested for its fine byssus, especially in the Gulf of Taranto and south Sardinia. Unfortunately, in the

1960s, divers have destroyed its populations for the reckless belief that they produced worth pearls. In fact, *P. nobilis* sometimes produces a small calcite pearl (Farn, 1986), reddish, white, or black whose commercial value is practically zero, and certainly does not justify the harvesting. In addition, containing a high percentage of water, these pearls dehydrate over time and crack easily. Today, *P. nobilis* is a protected species throughout the Mediterranean, and its populations are recovering and have once again become common in many places.

Even some gastropods can sometimes produce pearls often with iridescent colours, from yellow to blue: in general are not very popular with jewellers, but in some cases, especially if huge, have become famous. The best known are those produced by abalones (*Haliotis rubra, H. iris, H. laevigata*) from the coasts of California, Australia, and New Zealand. These are gastropods which are also farmed for food. These pearls have characteristic iridescent colours, ranging from deep green to red and blue. One of these, the large baroque "Big Pink", weighing 470 carats, was found in the waters of Petaluma, California, in 1990 and evaluated more than 4 million dollars.

The pearls produced by *Melo melo* are yellow-orange-golden and often large (even over 200 carats); this is a gastropod relatively common in Malaysian waters and in the South China Sea. Even the "Queen conch," *Strombus gigas*, a species widely distributed in the Caribbean Sea and abundantly fished for its meat, sometimes produces pearls, not nacreous, variable in colour, often pinkish, called "conch" which, if large, are really appreciated by jewellers.

On the contrary, cephalopods rarely produce pearls: it is possible to find them, generally yellowish, in the nautilus (*Nautilus pompilius*) from Australian seas (Farn, 1986).

Pearls in history and legend

Since ancient times, the pearl is linked to love, but at the same time, to pain: according to an ancient Eastern tradition, Adam's and Eve's tears for the death of their son Abel formed a lake where the pearls were born, from love and pain. Indeed, the association between tears and pearl is a concept still very present in the collective imagination and according to popular belief, pearls lead not only love, but also to tears for their owners.

Perhaps it is for this legend that some owners have become so popular: Pliny the Elder tells us that Queen Cleopatra made a bet with Mark Antony (83-30 BC): she was able to spent 10 million sesterces only for a dinner. To win the bet, Cleopatra dissolved one of the most beautiful pearls never seen before in the vinegar, valued at 8 million sesterces, and drank it. Perhaps this story is just slander put about by Octavian Augustus who had a vested interest in discrediting Cleopatra, but it is certain that the love with Mark Antony ended up in blood. By order of Octavian, a second pearl, which belonged to Cleopatra, was cut in two to make the earrings to adorn the statue of Aphrodite in the Pantheon. Curiously, the same scene was repeated, centuries later, at the Court of England: Sir Thomas Gresham (1519-1579), economic advisor to Elizabeth I, crumbled one pearl of great value, dissolved it in wine, and offered it to his Queen to make the startled ambassador of Spain understand who was the economic power of the Kingdom (Taburiaux, 1985). Even to Lollia Paulina, the third wife of the Emperor Caligula (12-41 AD), famous for her love for jewellery, pearls did not bring any good: she divorced after only a year of marriage. Another story mentioned in the *De vita Caesarum* by Suetonius (70-126 AD) reminds us that the same Caligula gave to his favourite horse a harness covered with gems and pearls, perhaps to spite his wife: but Caligula eventually came to a bad end. Good luck was not brought by the "Peregrina", one of the most famous pearls in the world and a wedding gift of Philip II (1527-1598), King of

Spain, to his wife Mary Tudor, Queen of England (1516-1558). Bloody Mary, as she was called later, died two years after her marriage, leaving no heirs. Even to Mary Stuart (1542-1587), Queen of France and Scotland and cousin of Elizabeth I, the pearls did not bring much good: during her captivity, she gave a necklace of 34 freshwater River Tay pearls to Thomas Howard, fourth Duke of Norfolk, but the marriage never took place at the behest of Elizabeth. After the beheading, Elizabeth "inherited" the collection of pearls of her cousin, but, in this case, we cannot say that they brought bad luck; indeed, many portraits of Elizabeth I boast an incredible amount of pearl necklaces. For another Elisabeth, things went very differently: according to legend, Empress Elisabeth of Austria (1837-1898), the famous and unhappy Sissi, never separated from a beautiful pearl necklace. During a staying in her villa at Corfu, she complained that the pearls had lost the original shine and was persuaded to leave the necklace in the sea: it was common belief that when in contact with sea water pearls could regain their original splendour. Little time later, the Empress was murdered in Geneva and the necklace, of course, was never seen again. Even the Empress Marie Eugénie (1826-1920), wife of Napoleon III, was very fond of pearls: she suffered a serious outrage, and after the defeat of Sedan (1870) and the collapse of the French Empire, fled to England helped by an American dentist, Thomas Evans, probably an agent of the American secret services. Out of gratitude, the Empress, in exile, gave him some gems that would be donated to the Pennsylvania Dental School, founded by Evans himself. Some of these now form part of the collections of the *Cooper-Hewitt Museum* in New York.

Figure 51: Portrait of Maria Theresa, Empress of Austria (1717-1780), by an anonymous painter, showing the huge use of pearls (probably fake) adorning the ceremonial dress.

Historical pearls

The Pearl of Asia*: the largest known baroque pearl. Egg-shaped, weighing 575 carats, and measured 76 x 50 x 28 mm. It was probably found in the Persian Gulf and belonged to the Emperor of China, Emperor Shah Jahan (1628-1658), and later to the King of Persia, Nadir Shah (1688-1747), and then still another Chinese Emperor, Qianlong (1711-1799). It is inside a cabochon mounted in a gold leaves and branches, decorated with jade and rose quartz.*

The Southern Cross*: the most popular Australian pearls. It is a set of nine pearls, seven in the vertical and two in the horizontal axes, all the same size and fused to form a cross. Found in 1883 in Baldwin Creek waters, it was sold for £ 10. Currently, it forms part of the crown jewels of England.*

The Georgibus*: 126 carats baroque pear-shaped pearl. Probably found in India, it was sold in 1620 by the merchant Francois Georgibus from Calais to Philip IV of Spain, who had it put on his hat. It was then considered the largest pearl ever seen in Europe.*

The Hope*: baroque bronze in colour, pear-shaped pearl, 425 carats and approximately 8 cm long. It owes its name to its first owner, the London banker Henry Philip Hope. Considered one of the largest pearls ever found, it was sold for $ 20,000 in 1975.*

The "Huerfana"*: called "orphan" because fished in the Gulf of Panama but not inside an oyster. Perfectly shaped and large, it belonged to Inés de Bobadilla (? -1543), Governor of Cuba. It is now part of the treasure of the Spanish Crown.*

The Pilgrim or Zozima*: 28 carats of perfectly spherical silver pearl, probably by* Pinctada mazatlanica. *Part of the treasure of the Spanish crown, it was donated by Philip IV to his daughter Maria Theresa of Spain on the occasion of her marriage to Louis XIV in 1660. Later it belonged to Youssoupoff, Princess of Russia (1769-1841), who had it made into an earring by Zozima, the Greek jeweller of the Tsar. In 1987, it was sold at Christies in Geneva for more than $ 460,000.*

The "Peregrina" or The Wanderer*: one of the most famous pearls of all times which is intertwined with 500 years of European history. 50 carats and* *pear-shaped, it was found around 1515 in a* Pinctada mazatlanica *in the Gulf of Panama and donated by Philip II, King of Spain (1527-1598) to his wife, Mary Tudor of England in 1554. Then it belonged to Queen Isabella Clara Eugenia of Austria (1566-1633) daughter of Philip II, sovereign of the Netherlands from 1598 to 1621 and to Queen Margherita of Spain (1584-1611), wife of Philip III (1578-1621). Diego Velazquez (1599-1660) portrayed both the Queens with this pearl. Forming part of the treasure of the Spanish crown, it fell into the hands of Joseph Bonaparte, Napoleon's brother and King of Spain from 1808 to 1813 and after to Ortensia de Beauharnais, adopted daughter of Napoleon I. She sold it to James Hamilton, Duke of Abercorn, to finance the political ambitions of her son Charles Louis Napoleon (1808-1873), the future Napoleon III, emperor of France. In 1969, the cinema actor Richard Burton bought it for $ 37,000 to give to Elisabeth Taylor on the occasion of her birthday. Mounted by Cartier on a necklace of rubies, diamonds and pearls, it was sold at auction in December 2011 by Christies in New York for the incredible price of $ 11.8 million.*

The "Regente": *as big as a pigeon's egg and weighing 86 carats, it is one of the largest and most perfect pearls known. It was bought by Napoleon Bonaparte in 1811 to decorate the imperial crown of his second wife, Marie Louise of Austria. Afterwards, Napoleon III gave it to his wife, the Empress Marie Eugénie. It was purchased at auction in 1887 by the jeweller Peter Carl Fabergé on behalf of Princess Youssoupoff of Russia. It sank into oblivion after the Bolshevik Revolution and reappeared in 1988, at Christies in Geneva, where was sold for more than $ 850,000.*

The Mary Mancini's Earrings: *drip beads belonging to the Medici family of Florence. When Maria de Medici married Henry IV in 1600, they became part of the treasury of the Royal House of France. Maria's daughter, Henrietta Maria, took them as a dowry to her husband, King Charles I of England. Then, Henrietta gave them to her grandson, Louis XIV who gave them to his mistress, Anna Maria Mancini, niece of the Cardinal Mazarin. In 1979, Christies sold them for $ 230,000.*

Freshwater pearls

Even some freshwater bivalves produce pearls, and sometimes of large economic value. In European, Asian, and North American freshwaters, several unionids, bivalves which can produce pearls sometimes of great value, live.

Figure 52: *Margaritifera margaritifera*, a pearly freshwater bivalve.

Their pearls are almost always irregular in shape and elongated silvery-white, rarely exceeding 6-7 mm. In Europe, these bivalves *(Margaritifera margaritifera* (Fig. **52**), *M. auricularia, Anodonta anatrina, A. cygnea, Unio pictorum, U. mancus, U. tumidus* and *Pseudanodonta complanata*) were once common in both rivers and lakes, with populations often numerically very important. Unfortunately, their excessive harvesting, which began in prehistoric times, and the subsequent pollution of the waters have destroyed many populations today. In fact, the entire group of freshwater bivalves suffering: *Margaritifera auricularia,* for example, is considered a highly endangered species, under extinction.

The Scottish freshwaters are, or rather were, rich in unionids and in particular *M. margaritifera*, a species famous for its pearls. Several of them are of such importance as to adorn the Royal Crown of Scotland, the so called "Kellie", found in a river near Aberdeenshire in 1621. Famous is also the "Abernethy" or the "Little Willie", an 11 carat pink, spherical pearl, with a diameter of 12 mm. It owes its name to the last Scots pearl fisherman, William Abernethy, who found it in the river Tay in 1967.

In North America, approximately 300 species of freshwater bivalves are present, many of which are virtually extinct *(Lampsilis higginsii, Hemistena lata, Epioblasma haysiana).* In Virginia, in the Mississippi valley and in the rivers of Arkansas and Tennessee famous pearls were harvested: the hope of finding a pearl unfortunately caused a rapid destruction of the natural populations of these bivalves, collected by their millions. In 1857, Jacob Quackenbush found a pink pearl weighing 23 carats in Notch Brook, New Jersey: purchased by Charles Tiffany for nearly $ 1,500, it was sold, probably thanks to Cartier, to Marie Eugénie, Empress of France, becoming the "Queen Pearl". Other famous American pearls are "the John", 17 mm in diameter, cream-coloured with pink reflections, was found in the Black River, Arkansas, in 1966, the "Venus", a 19 carat pink pearl measuring 17 mm x 11 mm, and all those composing the "Labor of Love" necklace, pearls of 3-11 mm diameter, for a total weight of 128 carats, all found in Tennessee River. Still, the "Pearl of Many Colours", found in the Cumberland River, shows extraordinary colours, a set of blue, green and pink pearls, probably due to the fact

that the mollusc lived in polluted water, and their nacre contaminated by chromium. It weighs 38 carats and measures 24 mm x 22 mm x 16 mm.
In China, the freshwater bivalve *Cristaria plicata* produces pearls of some interest and is also used in farming. Famous pink pearls were also produced by freshwater bivalves in Bangladesh, but today, following the pollution of the rivers, the production is much over.

The huge calcite pearls

Only when the calcium carbonate is deposited in lamellar and concentric layers of aragonite, the pearl becomes a real gem. Not all the pearls are, in fact, nacreous: large clams, for example, can produce huge opaque "pearls" in calcite, which being rare can still have a high value. They are the “pearls” that Captain Nemo cultivated in his submarine cave according to Jules Verne: *I saw a loose pearl, whose size equalled that of a coconut. Its globular shape, perfect clearness, and admirable lustre made it altogether a jewel of inestimable value. Carried away by my curiosity, I stretched out my hand to seize it, weigh it, and touch it; but the Captain stopped me, made a sign of refusal, and quickly withdrew his dagger, and the two shells closed suddenly.*
Up-to-date, the largest "pearl" of the world, more than 6 kg of weight for a size of approximately 24 x 14 cm, is the “Pearl of Lao-tze” or the “Pearl of Allah”, found in Palawan, Philippines, in 1934: it is a calcite concretion produced by *Tridacna gigas*. Donated by a village chief to Wilburn D. Cobb to thank him having cured his child of malaria, it is now owned by a jeweller in Beverly Hills, California. It was valued nearly 40 million dollars. A second pearl like this, but smaller, than the “Palawan Princess”, weighing more than 2 kg and measuring 15 cm wide, was auctioned by Bonhams in Los Angeles on 6 December 2009 at a base price of 275,000 euro (Fig. **53**).

Figure 53: A *Tridacna* shell with his huge calcite pearl.

Pearl fisheries

The areas of great historical tradition for pearl fishing (Fig. **49**) are the Persian Gulf, where harvesting was already active in the time of Alexander the Great (356-323 BC,) and the Red Sea from the time of the Ptolemy (300-50 BC). Along its coasts, from Jeddah to the Farasan Islands, at Massawa, Ethiopia, the pearl oyster was *Pinctada radiata*. Other important historical fishing areas are the Gulf of Mannar, South India, Sri Lanka, Madagascar, and the Malay Peninsula. In particular, the techniques of fishing and trade in the Gulf of Mannar were described by Marco Polo, in 1293, on his return from China. Until the modern age, the most important commercial centres were in India, mainly because Indian traders financed pearl fishing along the Arabian Peninsula and the Red Sea (Sharabati,

1981): in fact, once collected, pearls were sent to Bombay and Madras to be drilled and subsequently reached the major markets of London, Amsterdam, and Paris.

Between the 16th and 17th centuries, with the discovery of the New World, pearls of great beauty and importance, gems that were especially appreciated by the Incas people, suddenly appeared on the Spanish market. The Conquistadores not only quickly raided all pearls harvested by the local population, but also began a systematic withdrawal and devastating of the pearl oyster beds in the waters of the Gulf of Mexico, the Gulf of Panama, and along the coast of Venezuela, and in particular on the island of Margarita (meaning pearl, in Latin). The beds of the New World would be quickly destroyed in a few decades in the Atlantic as well the Pacific and Panamanian California, where Conquistadores discovered rich beds of *Pinctada mazatlanica*. The same would happen, following the great European explorations of the 18th and 19th centuries in the Pacific Ocean, and in particular, from the north-western coast of Australia to the Philippines, the islands of French Polynesia and New Caledonia. In a few years, the Pacific would become a major area of production of pearls.

In any part of the world the pearl fishing occurred, it has been a story of destruction: in a very short time the Europeans destroyed all the known beds. The discovery of pearl banks in the tropics provoked, in fact, a kind of "gold rush" with a devastating exploitation not only of the banks themselves, but of the natives themselves. Also those of French Polynesia, where *Pinctada margaritifera* of the Gambier Archipelago produces the black pearl, suffered severe damage when in 1842 the French laid the Tuamotu Islands under a protectorate, beginning the exploitation of the natural reefs: early in the 20th century it lead to their almost complete destruction. Suffering the same fate banks in north-western Australia, Torres Strait, and those around the cities of Darwin and Broome, where, by the end of the

19th century, millions of pearl oysters from *Pinctada maxima* were collected using Aborigines, Chinese, and Malaysians, reduced practically to slavery, forced to work in inhumane living conditions (Bailey, 2001). Fishing in apnea on the banks, at 20-30 m depths, was in fact dangerous and exhausting work, and not just because of the presence of sharks. In fact, the discovery of a beautiful pearl could arouse very dangerous interests from the other fishermen taking into account that it was always a rare event. In fact to find a good quality pearl, it is necessary to open around 500 oysters in the Persian Gulf, and 5,000 in the Philippines. In Polynesia, 15,000 *Pinctada margaritifera* oysters have to be open to find a black pearl (Doumange and Toulemont, 1992). The lunatic exploitation of the natural beds caused the collapse of the production: in the first decades of the 20th century, the fishing was finished in the Pacific, including the concomitant harvesting of oysters by the Japanese who, in the 20's and 30's of the last century, had to supply their industrial cultivation of pearl oysters. In the same years, a serious crisis also struck the Persian Gulf, the largest area of production of the 19th century, where 30,000 men worked on over 4,000 vessels (Sharabati, 1981). There are two main reasons for the crisis: the discovery and exploitation of oil, which created a labour market much less dangerous, and especially the depreciation of natural pearls in the world market caused by the invasion of the Japanese cultured pearls, much less expensive, which took place at the beginning of 20th century. The commercial impact of the Japanese cultured pearls was very devastating, taking into account that it is very difficult to recognize a natural pearl from a cultivated one: it is only possible to distinguish them with X-rays, because the core around which a cultured pearl grows is very large. Even today, the Bahrain prohibits the import of cultured pearls, even if pearl fishing in that sea is done as a hobby or just to entertain the tourists.

Additionally, the Government of Western Australia sought, with the *Pearling Act 1922*, to protect the fishing activities on the natural beds, prohibiting the manufacture, sale, and possession of cultured pearls. The only tangible result was giving the Japanese the world monopoly of cultured pearls for decades: only in 1956, in fact, the first Japanese-American-Australian company was initiated for the production of the *White South Sea Pearl*, the famous Australian cultured pearl.

Even the natural beds of pearl oysters in European freshwaters (mainly in Scotland) and North American were quickly destroyed, not only by fishermen: river pollution, linked to the industrial revolution, reduced many populations, and some species were on the edge of extinction, but are now protected by the *Convention on International Trade of endangered Species* (CITES), the organization of the *UN Agency for the Environment* (UNEP). In North America, in addition, another factor has certainly contributed to the crisis of these populations: the harvesting to produce pearl buttons or triggers for the pearl farming. In particular, the unionid *Pleurobema cordatum*, the "Ohio Pig-toe", will be collected by the tons in the 20th century to be sent in form of seeds to Japan.

The cultured pearl

The natural pearl is a rare gem, replaced today by the cultured one: every year millions of cultured pearls enter the market, mainly cultivated in Japan, China, Australia, and Polynesia. This pearl, although is the natural product of an oyster, is due to human intervention that fosters development. Already in the 13-14th centuries, the Chinese obtained pearls by introducing an artificial body as an irritant in freshwater bivalves. According to a legend, Ye-Yi Yang began to introduce some porcelain or bronze Buddha figurines, creating real miniatures of mother-of-pearl to be sold to the faithful. In Europe, the great Swedish naturalist Karl von Linnaeus (1707-1778) obtained freshwater cultured pearls as early as 1761 (Farn,

1986). However, the history of the cultured pearl is, by definition, a Japanese history. After the first experiments by Kakichi Mitsukuri, professor of zoology at the University of Tokyo in 1890, the son of a modest producer of noodles, Kokichi Mikimoto (1858-1954), after many attempts and many failures, with a delicate technique kept secret for years, realized the first farmed pearl. He introduced a small amount of the mantle (the tissue that produces the mother-of-pearl) taken from a second oyster, the "donor", and a small piece of mother-of-pearl, not into the mantle (as would seem more logical), but in the gonads of the "akoya" oyster, *Pinctada fucata.* Following the completion of this technique in the early decades of the 20th century, large plants were built in Japan. The commercial success was extraordinary and the Japanese cultured pearls flooded the markets around the world: only in 1941, Japan's entry into the war, something like 10 million pearls were produced and in 1952 the production was 31 million (Doumange and Toulemont, 1992). The farming takes place in sheltered bays, and up to the mid-60s of the last century, was controlled in all its phases by the famous “Amah” (in Japanese, daughters of the sea): the girls involved in the maintenance of the industry, who dive in the nude or wearing a traditional white robe. The farming of pearl oysters is a long and difficult process. After the transplant, it is necessary to wait at least three years for a pearl, which is not always successful: only two of how many transplanted oysters produce a pearl, and one out of every fifty, a perfect one.

After the Second World War, the Japanese techniques were copied by the Australians and large industries sprang up along the NW coast of Australia, where the "Southern White Sea pearl" by *Pinctada maxima,* was produced. Other farms have subsequently arisen in the Philippines, where a “golden pearl” highly appreciated is produced, as well as in Burma and Polynesia. Since 1961, in Bora Bora (Tahiti),

Pinctada margaritifera cumingi was farmed to produce the famous "Tahiti Black Pearl" (Doumange and Toulemont, 1992).

Commercial pearls are also produced by unionids, the freshwater bivalves. In the mid-50s of the last century, the United States began their farming, gradually reaching an excellent quality, albeit with some difficulty due to the complex life cycle that characterizes this bivalve. Other unionids, as *Hyriopsis schlegeli*, were farmed by Japanese in Biwa Lake. More recently, in the 90s, even the Chinese entered the market and it is not a coincidence that one of the major rivers of China is called the Pearl River. The Chinese are now producing large quantities of pearls, using *Cristaria plicata* and subsequently, with good results, also using *Hyriopsis cumingi*. Unfortunately, the final product of China is generally considered low quality and this has led to an inevitable fall in the market price of pearls. Europe is now experimenting with innovative systems to farm *Sinanodonta woodiana*, another unionid native from south-east Asia and Indochina, introduced in Europe in the early 1960s.

The value of a cultured pearl depends not only upon its size or colour, but also the thickness of the nacreous layer, of course with the same diameter. Generally the Australian pearls, the "White Southern Sea Pearls", of an average diameter of 12-14 mm and exceptionally up to 20 mm, have a thickness of the pearly layer that can reach 2-6 mm; those grown in Tahiti reach only approximately 2-3 mm. In contrast, the Japanese pearls "Akoya" rarely exceed a diameter of 8 mm, with a thickness of only 0.5 mm of the pearly layer. The economic value depends on the different productions, but the buyer has difficulties in identifying them. Even the colour, ranging from white to pink, purple to grey-black, is of importance and changing with fashion trends. Today, thanks to biotechnology, we can produce pearls with strange colours such as green, blue, and orange, mainly used in jewellery.

Fake pearls

There are fake pearls. Glass beads of various colours were already known in Ancient Egypt, as well as in Greece, in Etruria, and in Rome, between the republican and imperial period. In Venice, "the city of the blown glass", the market for fake pearls became prosperous in the 16th century. The Murano glass makers began to produce balls of blown glass, known as the "Pearls of Venice" or "Margaritas", filled with a waxy substance that gave them a special iridescence. In those years, the "conteria" or the art of creating threads of glass beads to make necklaces, crowns, and other ornaments was born (Morazzoni and Pasquato, 1953). This name may derive from the Latin comptus *("ornate"), but according to other hypotheses, it is derived from "counting cash", because the coloured beads were widespread as a monetary chip in some regions. In France, according to tradition, a certain Mr. Jacquin invented a kind of paint in 1610, the "Oriental Essence", with which he painted blown glass beads, which, of course, were sold as real pearls (known as Pearl of Paris). This sticky essence, consisting of guanine (iridescent crystals obtained from the scales of a fish, generally bleaks), cellulose nitrate, and alcohol, once dried, gave a special iridescence that, in some ways, resembled that of the natural pearl (Taburiaux, 1985). Today herring scales are used: these techniques are still used to produce admittedly false pearls. The most famous are the pearls of Majorca (and their imitations, Majorica, Mallorca, Orquidea) whose core may also consist of plastic or porcelain. With some care, it is easy to distinguish them: fake pearls, being made of glass, are very smooth to the touch, as opposed to natural and cultured pearls.*

Bohemian beads are lumps of mother-of-pearl, cut and polished in order to make them look like irregular pearls, but in this case the layers of nacre are obviously not concentric, and thus can be easily recognized. They are mainly marketed in Russia and Bohemia. By contrast, the pearls of Rome are made with alabaster which is subjected to chemical baths to give them the characteristic iridescence of the pearl.

New products are marketed as pearl with fancy names (Angel, Sheba, Mikomo, Takara, Kobe, Nikko, Sumo, Fijii, Aloha): in this case, the spheres are subjected to subsequent baths in various plastic substances mixed with lead carbonate, mica, and titanium oxide to give them an iridescent colour. Another real scam is to "burnish" the Japanese cultured pearls, passing them for Tahitian black pearls which are much sought after on the market, irradiate them with gamma rays, and then treating them with a solution of silver nitrate and ammonia (Doumange and Toulemont, 1992).

The mother-of-pearl

The mother-of-pearl or nacre, the inner layer of the shell of many molluscs, displays a typical off-white with iridescent reflections ranging from green to blue, pink to grey, depending on the species of mollusc: how do the pearls, on the other hand, have the same source. The nacre forms a layer particularly abundant in many molluscs, especially tropical species and in particular in trochids and in pteriids (the pearl oysters), *Haliotis*, and in unionids. The decorative use of their nacre is, in fact, very old, 7,000 years BC, widespread in the Middle East with the Babylonians and Sumerians. The first Book of Esther says that in the Ahasuerus Palace in Shushan, the winter residence of the Persian Kings, there were "couches of gold and silver on a mosaic pavement of porphyry, marble, mother-of-pearl, and stones". The iridescence of mother-of-pearl makes it particularly suited to inlay and the production of small objects: the area of Ur in Mesopotamia 2,500 years BC and in particular to Beidha in the south of Jordan, a major trading centre along the main road that connected the Arabian Peninsula to the Jordan Valley and the Mediterranean (Sharabati, 1981). Its use was naturally more prevalent in Asian cultures, close to the areas of production: Chinese and Japanese people have always used and continue to use it, but medallions finished with mother-of-pearl are present in prehistoric sites in North America, especially in the south eastern United States, today's Florida, the territory of the native Seminole (Whiteford, 2002).

In the Middle Ages, its use spread to Europe, where it was employed for small inlays on religious subjects. Especially in the 18th and 19th centuries, mother-of-pearl inlays will be worked into all types of furniture and to decorate musical instruments, trays, handles of guns and swords, to make lampshades, hair clips and other items for the ladies (Fig. **54**), handles for cutlery and paper cutters, knobs for umbrellas and walking sticks, rosaries, fans (Fig. **56**), jewellery, and opera glasses. Mother-of pearl gaming counters (Fig. **55**) were first produced in Canton,

China early in the 18th century; by 1750 they were produced in very large numbers which continued until approximately 1840 (Neal, 2007).

Figure 54: A mother-of-pearl and red coral earring (Naples, Italy, 19th century).

Figure 55: A Chinese mother-of-pearl gaming counter (18th century). An incredible amount of gaming counters were fish-shaped: it has been thought that the original order, using the French word "fiches" (meaning counters), was misinterpreted by the Chinese with the English word "fishes".

Figure 56: A nacreous fan from Far East, 19th century (left). A silver mounted unionid shell used as a pill box, 19th century (right).

Mother-of-pearl buttons

The most common use of nacre is the button, achieved by working shells of different gastropods belonging to trochid family, especially but not only *Trochus niloticus,* and turbinids (*Turbo petholatus* and *T. marmoratus*), plus the unionids and the pearl oysters, the pteriids; in fact, from all molluscs that produce shells with a considerable thickness of nacre. Some industries use "paua shells" (*Haliotis iris,*

H. australis, H. virginea) from New Zealand, which has a rather limited thickness, to create more original buttons with special iridescence from green to blue. Today, the "paua" mother-of-pearl is mainly used in costume jewellery.

The use of mother-of-pearl buttons in Europe dates back to the late Middle Ages, prevailing at the Court of Louis XIV in the late 17th century, and reached its peak in the 19th century.

Some traditions have ancient roots: today again, some Londoners used to wear clothes completely covered with thousands of pearl buttons, called the *Pearly Kings and Queens*. Particularly well known is the *Pearly Harvest Festival* which takes place in October, at St. Martin in-the-Fields, near Charing Cross, where you can meet these somewhat quirky characters. This tradition goes back to the Roman domination, when Romans, formed an alliance with a tribe who lived near Tilbury, along the Thames, to conquer Britain, who used mother-of-pearl ornaments. As a reward for aid received, the Romans agreed to their allies not only the privilege of preserving their rights, but also to bring their mother-of-pearl ornaments as a sign of distinction. During great occasions, Londoners, the descendants of that population, still wear mother- of-pearl buttons.

The first factory of pearl buttons was instigated in Birmingham in 1825: using a rotary punch, the shells were perforated, creating many disks of a desired and consistent diameter, and then planed, drilled, and polished. In the 19th century, the pearl shell trade to produce buttons became impressive: 2,000 tons were imported into England in 1859 from all over the world (Saul, 1974). At the end of the 19th century in the United States, the Great Lakes, and along the Mississippi, Illinois, and Rock rivers, thousands of tons of freshwater bivalves were harvested and used to produce buttons. Only the Illinois River produced more than 400 tons in 1930 (Wood, 1931) to feed nearly 200 factories that had arisen over time along the river and its tributaries.

Gradually, the use of mother-of-pearl will fall into disuse following the industrial production of plastic buttons, much cheaper, but the prestige of the pearl buttons, which can never be imitated, has certainly not diminished. The pearl button is now a niche product and mollusc nacre industry is certainly decreased in the world: it is, however, trying to create an "ecolabel brand" not only to ensure the authenticity of the product (the nacre is, in fact, easily falsifiable), but also to offer a product that is ecologically and ethically compatible.

Cameos

With some shells, it is possible also to make extraordinary jewellery: cameos, engravings on stone or shell. The artist, using the differences in colour present, is able to obtain contrasts, isolating a figure in relief. The cameo can be engraved on any hard stone, usually quartz forms (such as carnelian, the Indian sardonyx, agate, and onyx) for which, depending on the colours of the layers, can be recognized as sardonic cameos from the dark brown background, or carnelian cameos, when the ground colour is more orange or reddish. The workmanship of these stones was already fashionable in the Classical Age, and Alexandria was one of the capitals of this particular art. The black helmet shell is one of the best for cameo-cutting. It has an onyx ground, a dark coat under the pale outer layer, so that the cut figure will stand out well. The inner lip of a large shell could yield several brooches. The two semi-precious stones named in the descriptions of cameo shells call to attention the fact that cameo-cutting, one of the finer arts of antiquity, used only stone as the base material, for centuries, before the lava came to be used for cheaper work. Shells were first used in 1820 in Italy (Fig. **57**).

In addition to semi-precious stones, you can also accomplish cameos by working some shells, whose use was already known in ancient times, in the Alexandrian period of the 2nd century BC. But, only in the Renaissance, when the possibility of importation of tropical sea shells suitable for engraving at the high level, shells would gradually replace the stones (Cossignani, 1992).

Only six of the more than one hundred thousand species of sea shells in the world, can be used in glyptic art, the art of influencing the stone, creating cameos: these are the so-called helmet conch (*Cassis madagascarensis, C. cornuta,* and *Cyprae-cassis rufa*), shells more suited to emphasize the incision due to their background of a very intense colour. Cameos are also obtained by working *Cyprea tigris*, *Strombus gigas,* and *Pinctada margaritifera. Cassis madagascarensis*, in spite of his specific name, a misunderstanding of the type locality by the original author comes from the Western Atlantic Ocean, and is known as the "sardonyx helmet".

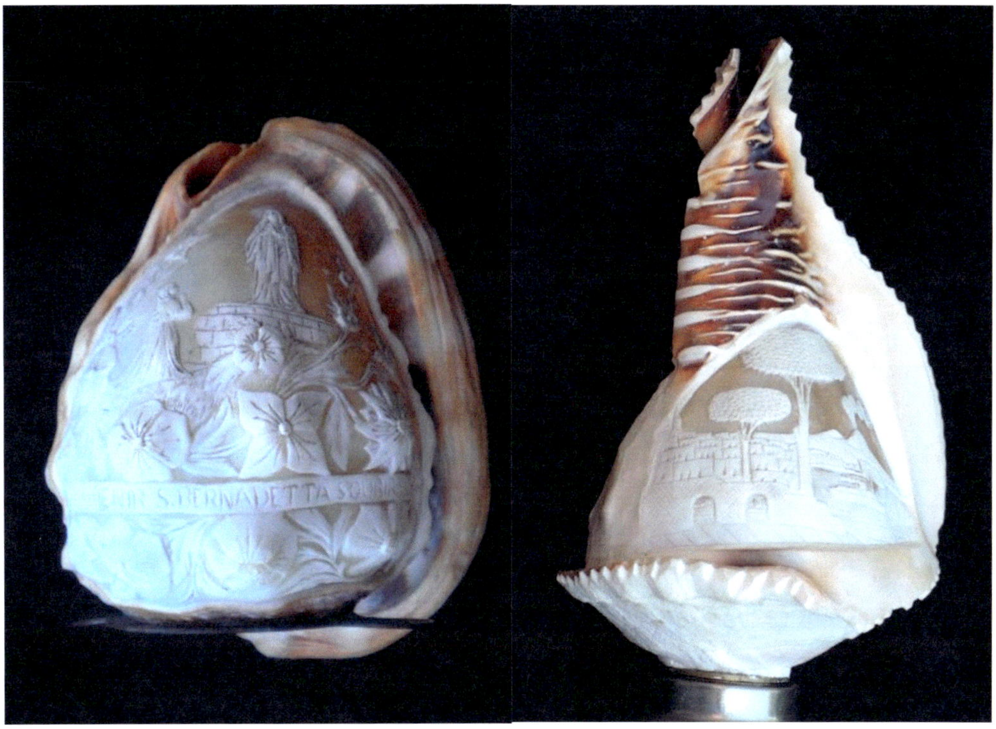

Figure 57: Shells carved as cameos (Torre del Greco, Naples). 20th century.

It is approximately 30 cm high, has a dark brown background and is perfectly white on the surface. *Cassis cornuta*, the "orange helmet", nearly 25 cm high, has instead the inner layer of orange colour and the outside white (Fig. **58**), while *Cypraecassis rufa*, the "cornelian helmet", around 15 cm high, has an inner layer reddish while the outside is pale pink. These last two species are of African origin and are abundantly fished in the waters of Zanzibar and Madagascar.

Over time, especially French and Italian artisans devoted themselves to this particular art in vogue in the Napoleonic and Neoclassical age, with subjects imitating classical themes, such as female figures of great beauty, or reproduction of flowers. Napoleon Bonaparte (1769-1821) was particularly fond of these jewels that he had begun to appreciate during the Italian Campaign. He founded in Paris, the Glyptic School, in full competition with the Italian artistry and, as often happens when a powerful man loves something, this immediately becomes the fashion: for the whole of the next century, the cameo will inevitably be part of the jewels of a lady. In Italy, the capital of the production was and remains the Torre del Greco, near Naples, where during the 18th century we find the skill returning in Naples, where Charles III Bourbon (1716-1788), King of the Two Sicilies, founded the *Royal Carving School*. Thanks to the munificence of this King, great bas-reliefs and splendid necklaces were created by artists attending that school in a social context of the highest artistic level and where, by tradition, hard stones were worked. Later (1878) the *Institute of Art and the School of Engraving on Coral and Design of Industrial Art* was born, to train young people in the art of engraving, teaching the processing of the cameos, to engraving of turtle shell, lava stones, red coral, and other semi-precious stones.

The extraordinary development of the manufacture of cameos in Torre del Greco coincided with two events seemingly far removed from each other: the arrival of large quantities of tropical shells that ships from Africa were using as ballast, and the discovery of rich subfossil red coral deposits in the Strait of Sicily, between 1875 and 1880, which caused a severe crisis, not only because it led to the saturation of the market, but for the decline of the quality. Only companies of greater tradition passed the crisis, developing a production of cameos of high artistic value. Famous among them is the *Allegory of the greatness of England on the sea*, a cameo on an entire shell, made by the Italian artisan Luigi Sabato over ten years. Throughout the world, especially in the Far East, cameos were also made using machines and pantographs, but the irregularity of the surface of the shell does not facilitate the machining that is used especially on pearl oyster shells (Fig. **59**, **60**). They are cheap products that have nothing to do with the handmade production in Torre del Greco.

Figure 58: A cameo from *Cassis cornuta*, the "orange helmet" (Naples).

Figure 59: A *Cypraea tigris* shell worked as a cameo using machines and pantographs (20th century).

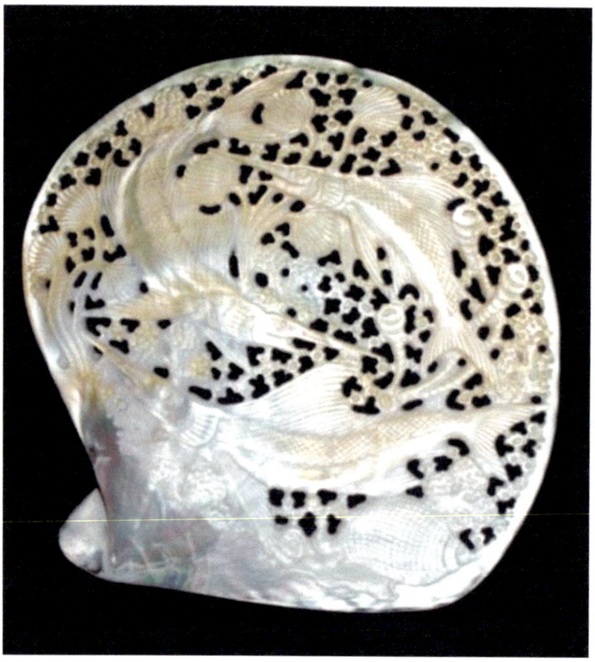

Figure 60: Worked nacreous layer of a *Pinctada* shell from Bali (Indonesia) (20th century).

CHAPTER 8

Dyes, Tissues, and Materials

Abstract: Some gastropods are capable of producing a pigment, the purple dye, which has been widely used in the past to dye a lot of different materials, while some bivalves can produce byssus fibres used in the past to make a very fine tissue. Cephalopods, octopus, and cuttlefish produce, thanks to specialized glands, a black pigment which is released in cases of danger to confuse the aggressor and promotes their escape: for centuries, it was used as ink, but is also appreciated in cooking pasta and rice.

In the Classical Age, Phoenicians were certainly the first and probably unique historic producers of purple, while in the Middle Ages, the pigment used at the Court of Charlemagne, was probably extracted by Britons from *Nucella lapillus*, a common gastropod from the North Sea. Little by little, this dye fell into disuse and around the 16th century, the dye shops in Europe were no longer be able to continue the ancient processes developed by the Phoenicians and the use of purple was lost forever.

Many bivalves, but overall, the Mediterranean *Pinna nobilis*, can produce silky filaments, the byssus, a set of keratin filaments which can be processed to realize rare and filmy tissues of a nice golden brown colour. For many centuries, the byssus tissues were a hallmark of command and power, and were reserved to noblemen and important members of the Church in Rome.

Keywords: *Bolinus,* byssus, *chlamys*, dyes, ephod, *Hexaplex,* muricids, noble pen shell, *Nucella, Ocenebra, Pinna nobilis,* purple room, royal purple, sea silk, red ammonite, *Thais.*

And they shall make the ephod of gold, of blue, and of purple, of scarlet, and fine twined linen, with cunning work.

Exodus, 28: 6

Riccardo Cattaneo-Vietti, Egidio Trainito and Mauro Doneddu

Since ancient times, Man learned to obtain not only a nutritious food, pearls, mother-of-pearl, or objects of daily use from molluscs, but he was also able to extract other valuable products. Some gastropods are, in fact, capable of producing a pigment, purple dye, which has been widely used in the past to dye materials, while a very fine tissue was made using the byssus fibres produced by some bivalves. In all cases, these products bear ancient sacred meanings.

As Exodus reminds us, in fact, the *ephod*, the sacred breastplate worn by Aaron, the brother of Moses and later the chief priest of the Jewish people, had to be woven with byssus threads and coloured with purple, a sign of great distinction. Today, purple and byssus threads are no longer used and only the *P'til Tekhelet*, the *Association for the Promotion and Distribution of Tekhelet* in Jerusalem, keeps alive this tradition of tissues dyed with purple, exclusively for religious purposes. Cephalopods, octopus, and cuttlefish produce, thanks to specialized glands, a black pigment which is released in cases of danger to confuse the aggressor and promotes their escape: for centuries, it was used as ink, but is also appreciated in cooking pasta and rice. In contrast, the Indian black ink, often confused with the squid ink, is a vegetable product. Squid ink was used for colouring tissues, especially by the Jewish people, and to produce a blackish brown ink, according to a user already present in the Cretan palaces of Mycenae, and utilised until the 19th century, when it was replaced once again by synthetic dyes. Today it is used also in a different way: in the culinary art, which uses it widely with rice, or in homeopathic medicine, where it could be effective for the treatment of certain diseases.

The royal purple

From time immemorial, people have tried to extract dyes from minerals, plants, fungi, insects, and molluscs. From the latter, one of the most striking cases is the royal purple, a dense violet pigment with a nauseating smell, produced for defensive purposes by the hypobranchial gland of some gastropods (largely, the muricids). Also some sea slugs such as *Aplysia*, the so-called "sea hares", if disturbed, can emit an more or less violet ink, a by-product of their metabolism, expelling the pigments of red algae on which they feed (MacColl *et al.*, 1990).

In the Classical Age and until the Middle Ages, the royal purple was used to dye linen, cotton, silk and wool, to colour ivory and wood, to write and decorate parchments, or used in painting, with the name of *purpurissum*.

The purple just extracted is colourless, but in sun light it goes from yellow to blue, to stabilize in a very distinctive purple colour. Its active ingredient is a very complex molecule, similar to indigo, the bluish vegetable dye, already known in Asia 4,000 years previously. For more than 5,000 years, the Mediterranean populations and in particular the Aegeans, have learned to collect purple from some Mediterranean muricids, such as *Bolinus brandaris* (Fig. **61**), *Hexaplex trunculus, Thais haemastoma,* and *Ocenebra erinaceus.* Later, during the Middle Ages (and perhaps until the 18th century) in Great Britain and Brittany, the purple was also extracted from the Atlantic dog winkle, *Nucella lapillus*, common in the North Atlantic intertidal zone.

Another dye-producing mollusc is *Janthina* (Fig. **62**), a genus of pelagic gastropods whose diameter does not exceed 5 cm, with a beautiful colour between lilac and violet. *Janthina* spends its life floating on the sea surface, attached to a raft of air bubbles that it produces. In past times, they could be so numerous that, after a storm, billions of them were washed ashore on the beaches of Florida but today, they have become much rarer.

Figure 61: The Mediterranean muricid, *Bolinus brandaris*, was commonly used by Phoenicians to extract the royal purple.

It produces a dense violet ink that it uses as an anaesthetic to paralyze its prey, usually jellyfish, including the poisonous Portuguese Man-o'War (*Physalia physalis*). It is very likely that this purple substance was used as a dye by the indigenous peoples of Florida.

Janthina is also present in the Mediterranean, but does not reach such quantities as to be able to be utilised. For some Jewish scholars, but not everyone agrees, *Janthina* could be the mysterious and rare "chilazon", the organism from which, according to the *Talmud*, was removed the blue pigment to colour the fringes ("tzitzit") present at each of the corners of the prayer shawl, the "Tallit".

The use of dyed material using pigments from gastropods was also common in cultures outside the Mediterranean: in the Caribbean area it is known that the people used pigments extracted from *Plicopurpura patula*, while along the American coast of the Pacific Ocean *Thais planospira* and *T. pacifica* were used. Other species are still used in China, Japan, and Peru.

Figure 62: *Janthina janthina*, a planktonic gastropod which produces a dense violet ink.

In the Mediterranean, the Phoenicians were certainly the first historic producers of the royal purple: they had an exclusive monopoly, and the cities of Tyre, Sidon and Lydda (today's Lod) were, for centuries, the main centres of its production and trade. It is no coincidence that the present-day Lebanon was known in antiquity as Canaan, which means "land of the purple", and the name Phoenicians comes from the Greek "phoenix" or "red," probably does not refer to their dark skin, but to the dye that made them rich and famous.

According to the myth, Melqart, the founder of the town of Tyre and the Phoenician equivalent of the Greek Heracles and Roman Hercules heroes, discovered the purple in making a gift to a nymph with whom he fell in love. Walking along the beach, the nymph Tyros particularly admired the colour of a mollusc and refused her love to the hero if he had not donated a dress with that particular colour.

Greek mythology incorporated and expanded the Phoenician story. Now Heracles discovers this pigment: always walking along the beach with his beloved, saw a dog bite a mollusc (a muricid, obviously) and gained a beautiful purple colour around its mouth. The girl said that she would never accept him as a lover, if he did not given her a dress of that colour. So Heracles gathered many molluscs, extracted the dye from them, and dyed the first dress of that colour, later called Tyrinian purple. This legend, also depicted by Pieter Paul Rubens (1577-1640), has some element of truth: *Thais haemastoma*, one of the main Mediterranean gastropods from which is possible to extract purple, can be found very often in late spring between the rocks in shallow waters, where a dog would certainly have been able to find it.

The great commercial expansion of the Phoenicians around the Mediterranean basin, between the 12th and the 9th centuries BC, probably was linked to the search for sites suitable for the production of salt, as well as the need to find new muricid populations to increase a certainly lucrative industry. The same Phoenician settlements in Sardinia are related to this activity, and in almost any Phoenician sites, incredible heaps of shells from which they drew the purple are still possible to encounter along the coast (Grimberg, 1968). The extraction of this precious pigment in the Roman period has been historically documented also in Taranto (southern Italy), in San Teodoro, Sardinia, and Essaouira, Morocco. To remember this old activity, a Moroccan monetary bill shows a *Bolinus brandaris*, the main muricid from which the pigment was extracted.

The process of obtaining the purple and the dyeing techniques were already described by Pliny the Elder (23?-79 AD). The fishing was carried out with traps: muricids are, in fact, greedy scavengers, and it is easy to attract them with appropriate bait. In addition, during the reproductive season, they have the habit of collecting in their hundreds, so it is extremely easy to harvest them. The exploitation of these populations was huge, considering that at least 8-12,000 specimens were needed to produce a gram of dye. Once caught and their shells broken the animals were pressed and left to macerate in large bowls for three days in the sun, mixed with sea salt. Subsequently, the "juice" obtained was boiled for ten days and foamed continuously, to clarify and condense it. The material was immersed for a time proportional to the intensity of colour that was intended, and finally exposed to the sun. It is easy to imagine the stench that had to enact these establishments: the Greek historian Strabo (58 BC-21? AD) wrote about the “unbreathable air” of the town of Tyre.

Different muricids gave different colours: purple from *Bolinus brandaris*, while the *Thais haemastoma* dye tends to red, and that of *Hexaplex trunculus* is more bluish: their combined use permitted more or less intense colours. Considering the high commercial value of this pigment, it is possible that the dye technique provided for a pre-treatment with plant pigments extracted from some lichens, such as the Atlantic *Roccella phycopsis* or the Mediterranean *Rytiphlaea tinctoria*: their use allowed, in fact, a saving of some dye from muricids and the final colour could also be deeper and more brilliant.

All ancient peoples, Jews and Greeks, Assyrians and Babylonians, Etruscans and Romans, until the Carolingians, made extensive use of purple dye; the *colour officialis*, being very expensive, was reserved for nobles and cardinals, princes of Church. The Roman tunic showed a single stripe of purple, the width of which pointed to the social role of the owner: the *angusticlavium* for the knights had a red band of 2-3 cm, that of *laticlavium*, reserved to senators, was 6-8 cm wide, while the *praetexta* toga, prerogative of the magistrates and priests, had a purple border with a width of at least 8 cm. In the Roman Republic and later Empire, consuls, victorious generals, and emperors also wore the *chlamys*, a short coat suitable for riding, dyed with purple. So, the purple dress was a symbol of power and immense wealth. It is not by chance that Luke (**16**: 19-31) compares the Pharisees to the rich Epulone *who was clothed in purple and byssus, and fared sumptuously every day*, while Mark (**15**: 21-32) points out that, to mock Jesus, the King of the Jews, the Roman soldiers *dressed Him up in purple, and after twisting a crown of thorns, they put it on Him.*

During the First Roman Empire, the materials of purple colour became increasingly popular, especially among the prostitutes. Even John, when in Apocalypse (**17**) he curses Rome, the new Babylon, calling her mother of all harlots and abominations of the Earth, depicts her adorned in purple and scarlet: *and the woman was clothed roundabout with purple and scarlet, and gilded with gold, and precious stones, and pearls, having a golden cup in her hand, full of the abomination and filthiness of her fornication.*

Symbols of power and sovereignty the purple dress was protected by the authorities but some emperors, including Constantine, sought to limit their use. But only in the Byzantine Empire, its possession was forbidden by Theodosius II (401-450

AD), who decreed that purple was reserved only to Emperor and his family. In fact, *Porphyrogenitus* (generated inside the purple) was the title of rightful imperial descent, because he was born in the Purple Room of the Imperial Palace of Constantinople: to be born in that room was tantamount to being recognized as legitimate heirs of the Emperor (Morelli and Saulle, 1998). According to the Byzantine ceremonials, an exclusive prerogative of the Emperor was to write with purple ink, to highlight rank and magnificence. These decisions inevitably caused the decline of the purple dye, whose extraction will end completely in 638 AD, when the town of Tyre fell under Muslim rule.

In the Middle Ages, the purple was still used at the Court of Charlemagne (742-814), using probably the one produced by Britons, extracted from *Nucella lapillus*, a common gastropod from the North Sea. But, but little by little, this dye fell into disuse. A demand remained, until the Renaissance, from the Church of Rome: Pope Gregory VII (1025? -1085), the Pope who excommunicated the Emperor Henry IV, emphasized, among other things, in his *Dictatus papae* on March 1075: *Quod solus possit uti imperialibus insigniis...*, or that he alone had the right to wear the imperial insignia (in other words, the front crown, the Phrygian cap, a sort of white fur cap and, of course, the purple ...), while in 1464, Pope Paul II (1417-1471) asked the cardinals to wear purple, perhaps to remind them that they must always be ready to shed their blood to defend the Church of Christ. But the dye shops in Italy and in Europe around the 16th century were no longer be able to continue the ancient processes developed by the Phoenicians and the use of purple was lost forever in Europe.

The byssus threads

Many bivalves, such as arcids, mussels, anomiids, noble pen shells, scallops, dreissenids and unionids, can produce silky filaments, the byssus, a set of keratin filaments made of a protein very similar to wool, of approximately 50 μm in thickness, which hardens quickly in contact with water and is used to anchor the shell to the rock and in sandy bottoms. The noble pen shell, *Pinna nobilis*, the largest bivalve of the Mediterranean, produces byssus filaments which may reach 15-20 cm in length: they were perhaps already processed by the Cretans, Phoenicians, Chaldeans and Egyptians to realize filmy tissues of a nice golden brown colour, but historical sources about this activity are uncertain: the first documented mention is by Tertullian in the 2nd and 3rd centuries AD. The uncertainty is also due to the fact that with the term "byssus" was used for all soft, transparent, and delicate tissues, regardless of the material with which they were produced. A rare and very expensive tissue known as “sea silk”, “mist” or “wind” tissue, was produced with the byssus of *P. nobilis*. In Taranto (Apulia, Italy), one of the Mediterranean centres of this activity since at least the 16th century, it was instead called “lanapesce” or “lanapinna”; the so called “tarantine tissue” is a wool derived from muslin and not from the byssus, as is often believed (Cecere and Mellea, 2009).

For many centuries, the byssus tissues were a hallmark of command and power, and were reserved to noblemen and important members of the church in Rome. Also the richer crusaders used to buy these tissues for their ladies in the ports of embarkation and disembarkation for the Holy Land such as the Apulian harbours. Subsequently, it became a niche product to be exhibited as a curiosity in the Universal Exhibitions in London (1862) and Paris (1867), or given away to celebrities, queens and kings. In several photographs, Giuseppe Garibaldi (1807-1882), the Hero of Two Worlds, often appeared with a curious byssus hat: it was a gift, from the Taranto people. The emergence of silk will bring this very expensive product to a gradual but inevitable decline, also taking into account to the environmental damages due to the strong fishing effort on the species. For example, to make the wedding dress of Maria Pia of Savoy (1847-1911), the daughter of Vittorio Emanuele II, King of Italy, when she married Louis I, King of Portugal, in 1862, 2,700 shells were sacrificed! The bridegroom, known for his passion for oceanographic studies and marine biology, perhaps did not appreciate this havoc.

At the end of the 19th century, Giuseppe Basso Arnoux (1840-1919), a physician from Cagliari (Sardinia, Italy) convinced himself that massages with byssus gloves were particularly effective, and attempted to develop a semi-industrial production. For this purpose, he formed the *Byssus Ichnusa Society* in London (Fig. **63**), but did not get the results that he expected.

Even in the 60s and 70s of the 20th century in southern Italy byssus undergarments, ties, gloves, and socks were produced: handmade products of high value, now completely disappeared, since *P. nobilis* has become a protected species by the European Community and its harvesting is quite rightly prohibited.

Figure 63: The *Byssus Ichnusa Society* in Carloforte (Sardinia) stamp.

Today the last weaver of byssus, Mrs Chiara Vigo, lives in Sardinia and cut the byssus directly under water, thus avoiding killing the bivalve, as had been suggested by Cerruti (1937, 1939) who had also suggested possible farming in the Taranto Gulf. For pure passion, Chiara Vigo continues to spin it off and weave it as did her grandmother, to produce unique pieces, often donated to museums around the world. Theoretically, it is possible to produce tissues using byssus from other bivalves: for example, it has attempted in Spain, where, in the mussel farms, a remarkable production of byssus by these bivalves is available, but the results were modest because the product is of poor quality.

Building materials

At different times and in different environments, Man, after having eaten the contents, often used the shells as building material. As early as the Palaeolithic, shells were used in large quantities around the house to promote soil drainage, especially in areas close to ponds where water tended to stagnate.

Large shell heaps, sign of the presence of the most ancient human settlements, have been used for years in the Eastern United States as a source of building material, often to form the base of the first American coastal roads (Ristine, 2009): their destruction unfortunately contributed to the irreparable damage to Pre-Columbian archaeological studies of human settlements along the coast of America. The same was the case with the oyster and other shellfish consumed by millions in the metropolitan areas and in the Caribbean to the 19th century: they become a source of calcium carbonate at practically zero cost, to obtain lime, or used as ballast in ships. In the tropics, along with stony corals, shells are still used to make walls: a very particular use of the Queen conch (*Strombus gigas*) at Iles des Saintes (Guadeloupe) in the Caribbean is for the graves in the Cemetery of Terre de Haut, all decorated with this large shell (Sutty, 1984).

Some shells are used in horticulture to condition the soil, increasing the concentration of calcium carbonate and therefore making it less acid. The "shell bottoms", used in marine aquariums, have a similar use: this layer of shells, a calcareous material passing continuously in solution, acts as a buffer to avoid an increase in the acidity of the water. In fact, seawater is basic (pH between 8.2 and 8.4), but in a confined space such as an aquarium, the production of carbon dioxide can acidify the environment: a shell bed, or alternatively, coral crush, are the ideal solution to this problem.

The cuttlefish-bone is also an excellent dietary supplement for many animals, especially for birds and land snails, being a source of calcium carbonate necessary for their metabolic needs, and in particular for the production of eggs in birds and shells in molluscs. Even at an industrial level, shell powder is added to the feed of laying hens, to favour egg calcification.

Finally, we must not forget the rocks containing fossils of many molluscs, especially ammonites or nautiloids, cephalopods with a chambered shell which appeared in the Devonian, 300 million years ago. Slates of these rocks, often common in Europe, Africa and South America, were always used in buildings: a classic example is the Red Ammonite, whose quarries are spread throughout southern Europe, from Spain to Turkey. It was used in the past by Romans as building stone, whose we have a splendid example in the Arena in Verona (Italy). Today, its use is mainly ornamental, for floors of great aesthetic impact.

CHAPTER 9

Madly in Love with Shells

Abstract: The passion to collect shells arose in the Upper Palaeolithic, and attended all the cultural evolution of Man. The breakthrough will come in the 16th century with the beginning of the great geographical explorations: the first large shells from the South Seas and the Caribbean which came in Europe received an immediate and resounding success. Beautiful and rare, and thus highly sought after, they began to enrich the collections of *Naturalia* in the *Wunderkammern,* extraordinary places, filled with collections of a disparate and chaotic expression of richness which arose in all courts of Europe.

The mid-19th century is called the Golden Age of Conchology thanks to Hugh Cuming ,who began to collect shells, especially in Chile and Argentina, and after from Easter Island to Polynesia and Philippines. At his death, a huge collection of nearly 53,000 shells of 19,000 species remained, that the British Museum bought.

The shell collector has an unrealizable ambition: to hold all the shells of the world. Conscious of the impossibility of realizing this dream, he often specializes in one group (generally a family). Today, the great collectors particularly appreciate some families of gastropods, especially the cones and cowries, whose shells are very smooth, with porcelain surfaces. Other families much sought after are volutids, olivids, terebrids, mitrids, muricids, strombids, cassids, cimatids and harpids. Bivalves arouse less interest, although thorny oysters and scallops are, perhaps, the only two families really appreciated worldwide. Recently, the interest of collectors was also directed towards the shells of molluscs living in deep waters, below 1,000 meters depth, as the pleurotomariids, primitive gastropods characterized by a long slit that runs through the last whorl of the shell.

Even many fossil molluscs are appreciated: some ammonites arouse considerable interest and can achieve amazing prices at the auctions.

Keywords: Ashmolean Museum, Cabinets de Merveilles, cassids, Cavendish, cones, *Conus gloriamaris*, cimatids, cowries, Cuming, *Epitonium scalare*, fake shells, harpids, Hirohito, *Lyncina leucodon*, *Mirabilia*, mitrids, molluscs, muricids, *Naturalia*, olivids, scallops, strombids, terebrids, thorny oysters, volutids, Wunderkammern.

Riccardo Cattaneo-Vietti, Egidio Trainito and Mauro Doneddu

No mention shall be made of coral or of crystal; the price of wisdom is above pearls.

Job, **28**: 18

It is perhaps more fortunate destiny to have a taste for collecting shells, than to be born a millionaire!

Robert Louis Stevenson (1850-1894).

You are examining my shells, Professor? Unquestionably, they must be interesting to a naturalist; but for me they have a far greater charm, for I have collected them all with my own hand, and there is not a sea on the face of the globe which has escaped my researches.

Jules Verne (1828-1905), *20,000 Leagues Under the Sea* (1870).

Shells show often such a refined beauty as to arouse a sense of pleasure, a kind of aesthetic pleasure in the myriad of simple passionate collectors who, at all times and all over the world, are always looking for and preserving, sometimes in an almost manic way. To collect objects is human nature, but in some cases collecting has led to real pathological forms, a compulsive need to assert one's own narcissism, in the reflection of the creativity of others, and in the case of shells, in the creativity of Nature itself. Certainly an amount of vanity characterized some historical shell collectors: all these aspects have made this kind of collection a fascinating and probably unique story, full of characters and amazing adventures, a story told in *A history of shell collecting* by Peter Dance (1986).

This "strange" passion arose early, in the Upper Palaeolithic. Certainly before 30-40,000 years ago, someone picked up and carefully arranged some shells on a ledge in a cave, at Arcy-sur-Cure, in Burgundy. That they were not the remains of an ancient meal it is understood by the fact that these shells were already fossils when someone, fascinated by those forms, collected and saved them (Leroi-

Gourhan, 1961; Caron *et al.*, 2011). With this action, the act of collecting was born.

Among the ashes of Pompeii (Tiberi, 1879) a small shell collection, perhaps belonged to Pliny who had given a detailed description in his *Naturalis Historia*, also emerged: among different Mediterranean species, some shells as *Cypraea pantherina, C. erosa, Conus textile* and *Pinctada margaritifera,* certainly from the Red Sea, were present, one of the first known malacological collections. During the long medieval period, when the Church preached against the ephemeral and unnecessary as suggested by Job, the desire to collect objects and precious things died down: in Europe, the pleasure of owning rare items, often for a unique aesthetic sense, will reassert itself only in the 15th century.

The Wunderkammern

At the end of the Middle Ages, scientific interest and the desire to understand the complexity of the universe through the study of Nature was still far: it would be strongly affirmed only during the full Renaissance period. At the beginning, many people started to collect rare and unusual natural items only for a personal caprice or a form of narcissism, more than for the spirit of knowledge: generally to amaze guests and friends and to increase, as a final result, their egos. The birth of a real cultural interest and of a refined collector of natural and precious objects (*Naturalia and Mirabilia*) occurred only in the Renaissance. Even the young Leonardo da Vinci (1452-1519) began to collect the natural curiosities that he discovered in the countryside of his native Tuscany, including many *nichi* (shells): his purpose was, however, different. He wanted to understand the intimate design of Nature and tried to interpret it: the fossil shells so common in his homeland, the first models on which to exercise his art, but also the reason to begin to think, in scientific terms, of the origins of those fossils and to discuss the Universal Deluge (**Chapter 15**).

The breakthrough will come only in the 16th century with the beginning of the great geographical explorations in the West and East Indies: the first large shells, from the South Seas and the Caribbean which came in Europe received an immediate and resounding success. Beautiful and rare, and thus highly sought after, they began to enrich the collections of *Naturalia* that powerful people then gathered in *Cabinets de Merveilles*, *Kunst-* or *Wunderkammern* (Leonhard, 2007): extraordinary places, filled with collections of a disparate and chaotic expression of richness, but also of a refined intellectual stature. They were, in a sense, the first moment of the development of the museum concept where, ideally, Man attempt to describe the Universe, realizing the whole of its complexity. In two centuries, dozens of *Cabinet des Merveilles* arose in the courts of France, Italy, Germany, Portugal, Denmark, Sweden and Russia and in the castles and villas of half of Europe.

In Italy, the Medici family in Florence, the Este in Ferrara, the Sforza in Milan and the Gonzaga in Mantua, gathered the best of what was available in the field of natural curiosities as well as in jewellery, in silver and semiprecious stones. Between 1570 and 1572, the grand duke Francesco I de Medici (1541-1587) *Studiolo*, adjacent to the *Salone dei Cinquecento,* was realized by Giorgio Vasari in the Palazzo Vecchio, Florence, in collaboration with Jacopo Zucchi (1541? -1590) to collect his *Mirabilia*. The historian Vincenzo Borghini described it: *has to serve as a wardrobe of rare and precious things, and currency for art, jewels, medals, carved stones, cut crystal vases and such things, not too much greatness, pinned in their own closets, each in its kind.* The collection of the grand duke, a passionate alchemist, necromantic and passionate researcher of the Philosopher's Stone, has been lost and no one knows exactly what valuables it could hold, but shells would certainly not be missing. At the same time, the Archduke Ferdinand II of Tyrol (1529-1595), son of Emperor Ferdinand I, rebuilt the Castle of Ambras near Innsbruck, creating a *Wunderkammer* for his collection of arts and curios. At that time, the *Cabinet of Wonders* of Ferrante Imperato (1550?-1631), apothecary and naturalist, housed in Palazzo Gravina in Naples, was well known: an etching (Fig. **64**), published in 1599 in the *Naturalis Historia* by Ferrante Imperato himself, illustrates this famous *Cabinet.*

Figure 64: The *Cabinet of Wonders* by Ferrante Imperato in Naples (1599).

The collecting trend spread rapidly: the Italian Ulisse Aldrovandi (1522-1605), the founder of the modern natural history, formed an impressive collection, now largely preserved at the University of Bologna, called a "microcosm of nature" by the same author. Other "museums" were set up by Francesco Calzolari (1521? - 1606), later described by Andrea Chiocco in Verona in 1622, and that in Padua (1656) by Moscardo (Caprotti, 1985). In 1649, the French physician Pierre Borel (1620-1671), a passionate collector, published a book in which listed and described more than 200 *Cabinet* he had visited throughout Europe (de Panafieu, 2008).

The royal collections and the first scientific collectors

Until the mid 1600a, gentlemen gathered shells to amaze friends. The first shell collecting in a systematic manner and scientific way was probably due to John Tradescant (1570-1638), "the first Gardener" of the Duke of Buckingham and King Charles I. He collected natural curiosities in Russia as well as in Maghreb, and his son, John the Younger (1608-1662) visited the New World and brought many species back to England. “The Ark”, as their rich collection was known, will then be acquired by Elias Ashmole (1617-1692), collector and alchemist, and it will became the *Ashmolean Museum* at Oxford University, in 1683, the first scientific museum in England and perhaps the western world. Even at the Court of France, a *Cabinet du Roi* at the behest of Louis XIII, was created in 1635, a collection that, a century later, will become one of the major collections of the 18th century, under the able guide of Georges-Louis Leclerc, Comte Buffon (1707-1788), one of the greatest naturalists of his century. In England in the early 18th century, Sir Hans Sloane (1660-1753), a British physician and naturalist, Sir Isaac Newton's successor to the Presidency of the Royal Society, was another great shell collector. When he died, his collection was left to King George II, after a payment of £ 20,000 to his heirs: it become the first fundamental body of the collections of the British Museum, which is still the most extraordinary malacological collection of world.

Margaret Cavendish Bentinck, the second Duchess of Portland (1724-1785) was famous among the great collectors of her time: Horatio Walpole, 4th Earl of Orford (1717-1797) said of her: *few men have rivalled Margaret Cavendish in the mania of collecting, and perhaps no woman. In an age of great collectors, she rivalled the greatest.* The Duchess made shells very popular at the court of King George III, and gathered an impressive collection thanks to the many scientists and explorers with whom she collaborated, including Daniel Carl Solander, a naturalist aboard the HMS *Endeavour* of the famous Captain James Cook. When Cook (1728-1779), one of the greatest navigators of all times, returned from his first voyage to the Pacific Ocean in 1771, he brought an impressive collection of animals and plants collected by Joseph Banks and Daniel Carl Solander. Some pieces were sold at very interesting prices, and many to the same Duchess of Portland. Unfortunately, after her death, the collection was auctioned and dispersed. The interest generated by tropical shells in the English aristocracy pushed officers and sailors to collect shells everywhere travelling in the "south seas", also hoping to increase their income. The tradition continued for many centuries and no sailor was to come back from a journey without bringing shells from far-off lands. At that time, German and English gentlemen put together famous collections, while the Dutch, thanks to the East India Company's commercial activities, imported an incredible number of shells from Asia for the pleasure of the collectors. In Amsterdam, shops specializing in sea shells and rare natural objects arose, as that of the naturalist Levinus Vincent (1658-1727), author of the famous *Wondertooneel der Nature*, the catalogue of his collection published in Amsterdam in 1706. The wide availability of shells, used as decorative objects in the homes of wealthy merchants, will also leave an important mark in Flemish painting, from the 16th to the 18th century, emphasizing the passion for collections of wonders from distant lands.

Malacological collections continued to grow in Europe favoured by the love of scientific curiosities, typical of the culture of Enlightenment. Finally, Karl von Linnaeus (1707-1778), the great Swedish naturalist, suggested the development of thematic collections, and imposed the criterion "to associate similar things and separate different things", indicating the correct way to describe the Nature and to understand the complexity of the entire universe. This will be the mission of natural history museums worldwide, where malacological collections become systematically organises.

The 19th century's great collectors

Collecting shells, as an element of distinction, would spread even to the middle classes, but only in the mid-19th century, the Golden Age of Conchology. In the first half of this century, Hugh Cuming (1791-1865), an eccentric, rich English businessman, with trade to South America, became the largest collector of all times in only a few years. At first, he began to collect shells especially in Chile and Argentina, then quickly retired from business and, moved by an extraordinary interest in knowledge, armed his yacht, the *Discoverer,* and set off exploring the Pacific Ocean in 1827, from Easter Island to Polynesia. In eight months Cuming collected an impressive amount of shellfish, focusing, among the first, on pearl fishing in the waters of Marutea, French Polynesia. He dedicated a second trip to the Pacific coast of South America, from the Island of Chiloe to Mexico, the first to use and experience a dredge for collecting samples in deep waters. In 1831, Cuming returned to England and entrusted to G.B. Sowerby and W.J. Broderip the whole collected material to study, after described in the *Proceedings of the Zoo-*

logical Society something like 2-3,000 new species for science (not just shells) (Dance, 1986). A few years later (1836), Cuming decided to visit the Philippine Islands, hitherto virtually unexplored, where he picked up other 3,045 species of molluscs, of which 550 were terrestrial, as well as a multitude of other organisms. He also visited Singapore, Sumatra and the Malay Peninsula, always gathering natural material. Finally, in 1840, he returned to England. The breadth and richness of the Cuming's collections are amazing: only many years of work by most famous European malacologists to whom Cuming distributed his material, would lead to the classification of all the collected material. At his death, a huge collection of nearly 53,000 shells of 19,000 species remained that, a few years later, the British Museum bought for £ 6,000.

The fake shells

The passion for collecting infected many people and some rare shells, either alleged or unavailable, reached unbelievable prices. Auctions were organized and often the desire to please the customer and make easy money pushed rogue traders into real scams. This economic interest became so important that in Netherlands, in the 18th century, a new profession was born, the "shell doctor" (Dance, 1986). In fact, often the shells were found dead and stranded, with faded colours: with deft strokes of the brush they were brought back like new. Sometimes, skilled artisans altered the shape of common shells, making them rare or "unique" ... and the "naïve" collectors bought them. This happened not only in the Netherlands: in La Rochelle, France, in 1767, certain bright golden limpets were placed on the market and commanded a high price. They were in fact nothing more than the common Magellanic limpet (Nacella deaurata*), but cleaned of their periostracum. In the US, some knavish shell dealers removed the outer surface of the olive shells with acids, giving them quite a different colour pattern. The shell would now be palmed off on an inexperienced collector as a different species or a freak of nature. A famous example of malacological scams involved* Epitonium scalare*, a gastropod from the China Sea, especially beautiful and highly sought after during the 18th century. This shell attained exorbitant and fictitious values on the European market. Chinese traders, to meet the constant request from European collectors, began to make perfect models by moulding them from a paste made of rice* *flour: imitations, however, that should not be washed, otherwise they dissolved! Obviously, these imitations today are worth far more than the original.*

Figure 65: The first illustration of the Precious Wentletrap, *Epitonium scalare*, which appeared in the *Appendix* to Rumphius' *D'Amboinsche Rariteitkamer* (1705).

At the end of 18th century, the European nations began to organize important geographical expeditions, sending ships around the world whose stated purpose was to gather major scientific information, but with less obvious, but certainly deeper, political, military and commercial reasons. Firstly, the Royal Navy: the scientific societies and the national museums of natural history in the meantime established in London, Paris, Vienna and Washington, were the charge of the collections and the specimens remained their property. These great explorations, which lasted throughout the 19th century and into the 20th, lead to the knowledge of the great biological richness of the Earth, but they gave some grief to collectors: a species, considered rare and paid for at incredible prices, suddenly became common, as new populations were found. At the same time, at the end of the 19th century, with the construction of the first steam ships it become possible to dredge even deeper, exploring an environment which until a few years previously was completely unknown: the deep sea become a valuable source of new species for collectors and scholars, and the latter now become professionals, working in large public scientific institutions.

The disappointments of rarity

Taste, fashion, and unexpected discoveries played, over time, many jokes on the collectors: a shell considered rare in the 1700s can suddenly turn out to be common in 1800, giving many troubles to the owner.

The concept of rarity in Nature is, in fact, complex and very "dangerous" for a collector: but a shell must have a price of affection, very different from that of the market ... What is rare today may become common in future and vice-versa. In nature, in fact, a species may be rare for a complex set of reasons. It is often considered rare because it has a short life cycle or its distribution area is very limited and we do not know exactly the environment in which it lives: maybe, it is an epibiont on another ignored species, or lives in environments difficult to access, with limited or inaccessible distribution, such as the caves. Once the ecological needs of the species are better known, it will be easier to find it. Other species are, however, objectively rare for their poor ability to compete with other species, which more efficient in exploiting a limited resource (this condition preludes to extinction) or, on the contrary, they are from the onset a phenomena of speciation that have not yet led to a complete quantitative level of the individuals bearing the new selected genotype. From this point of view, it is important to emphasize that in many species, local populations can show some recessive traits (especially chromatic, for example, melanistic or albino) expressed in different ways. In this case, the variety is rare and it will have a significant value to collectors, very different from the common form.

Finally, we must take into account that we might mistakenly consider a species as a hybrid or the result of the reproduction of two separate species, genetically close. It is a phenomenon that occurs rarely in nature, so when we meet an individual showing mixed characters of two species, we consider it a “true” species, while in reality it is only a "reproductive failure". In the case of a mollusc, its shell will inevitably be rare and therefore highly valued by collectors.

The story of *Epitonium scalare (*Figs. **65; 69**), the beautiful and elegant gastropod from the China Sea that was much sought after in the 18th century, tells this story true. The Duchess of Portland paid a fortune for it, and the Emperor Francis I (1708-1765), husband of Maria Theresa of Habsburg, paid 4,000 florins to obtain a specimen. Other shells of the same species were included in the collections of Catherine of Russia and the Queen of Sweden (Saul, 1974), a real rarity indeed. But already in the first years of 1800, its price had halved as a result of geographical explorations and discoveries of new populations.

Other shells became very popular, such as *Malleus malleus* and *M. albus,* bivalves from the Pacific with a curious hammer shape: at the end of the 18th century, they reached incredible prices, but today are worth only a few Euros. Even recently, the market has played considerable jokes: in the last thirty years, the commercial value of the Indo-Pacific cowry *Perisserosa guttata* decreased from $ 2,200 to $ 50. But there can also be another way: because of reckless collecting, some beautiful shells have now become rare, as in the case of an Australian cowry, *Zoila perlae*: in thirty years its market price has increased from $ 400 to $ 4,000.

The Mediterranean Sea does not offer the richness of the tropical seas, but contains some fine rare shell, sought after by collectors. *Mitra zonata*, a good-sized gastropod reaching 10 cm in height, is one of the most desirable. It lives in deep waters and shows a colour pattern in which transverse brown bands are alternated with lighter bands. Another species, *Ampulla priamus*, an Atlanto-Mediterranean species, pinkish in colour with dark spots, is much appreciated. Among the cowries, some are fascinating because of their rarity: *Schilderia achatidea* is a much sought-after species, slightly pear-shaped, and with the back a dirty yellow sprinkled with irregular brown spots. The base is white and the opening has many fine teeth that distinguish it from another cowry prized by collectors, *Zonaria pyrum* with coarser and more spaced out teeth. Another very popular cowry is *Naria spurca*, with a background colour of yellow and orange dotted with white.

An impossible ambition

The shell collector has an unrealizable ambition: to hold all the shells of the world. Conscious of the impossibility of realizing this dream, he often specializes in one group (generally a family) or even only a genus, doing everything possible to have a collection which is as complete as possible. Even in this case is not easy to achieve this ambition: all the great families hold hundreds of species, some of them very rare, with very few known specimens, preserved in the main museums of the world. It is almost impossible that an individual might have all the species of a given family. Moreover today, the collection and trade of certain species is often prohibited or regulated by international law (**Chapter 14**) for which it becomes impossible to gain possession. Despite these obvious difficulties, malacological collecting is becoming increasingly popular. Today, as then, the great collectors particularly appreciate some families of gastropods, especially the cones and cowries, whose shells are very smooth, with porcelain surfaces. Other families much sought after are volutids, olivids, terebrids, mitrids, muricids, strombids, cassids, cimatids and harpids. Bivalves arouse less interest, although thorny oysters and scallops are, perhaps, the only two families really appreciated worldwide.

Recently, the interest of collectors was also directed towards the shells of molluscs living in very deep waters below 1,000 meters depth, inaccessible to even the most powerful vessels. Only with oceanographic equipment, inaccessible to the common people, can these species be obtained, more often than not especially beautiful from the aesthetic point of view, and certainly very rare.

Figure 66: *Mikadotrochus rumphii*, a primitive gastropod with a conispiral shell from Japan, particularly appreciated by Emperor Hirohito. To remember this imperial interest, the genus of this species means "The Emperor's spinning top".

Among these, pleurotomariids, a primitive gastropod family characterized by a long slit that runs through the last whorl of the shell, are particularly appreciated, as not more than 20 species are known. They are considered true living fossils: among the most sought after, *Bayerotrochus tangaroana,* described in 1982 from the deep waters of New Zealand. Pleurotomariids are the molluscs loved by Hirohito (1901-1989), the Emperor of Japan, a passionate marine biologist, and so they became a symbol of the Emperor himself (Fig. **66**). Their collection was forbidden to ordinary people until the 1940s, and those who had by chance found a specimen were required to offer it as a gift to the Emperor (Hanson, 2007).

Even many fossils of molluscs are appreciated and have an important market: some ammonites arouse considerable interest in the field of collecting, and can achieve amazing prices at the auctions.

The cowries

Cowries, for their elegance of shape, generally ovoid, their porcelain surface, and the variety of colours and patterns, are the favoured shells of collection. They have also enriched the more sumptuous traditional costumes of Africa and Asia, and are symbols of prestige, wealth and magical power in many continents (**Chapter 2**). More than 200 species are known, some of which very common, such as *Cypraea pantherina*, from the Red Sea, *C. tigris* which is present throughout the Indo-Pacific from Africa to Hawaii, and the delicate and elegant *Lyncina carneola* and *Mauritia mauritiana*. Also the well-known money cowries (*Monetaria monetaria* and *M. annulus*) (**Chapter 6**) are very common in the Indo-Pacific.

Other species are very rare, and a long list of cowries is much sought after by collectors: among these, *Austrasiatica deforgesi* and *Nesiocypraea aenigma,* of which very few examples are known, while *Zoila (Thersites) contraria*, from Western Australia, may reach a value of 12,000 Euro (Fig. **67**).

Until the 1960s, the British Museum of Natural History had the unique specimen of *Lyncina leucodon*, the cowry characterized by white teeth, fished in 1828 in South Africa. Nearly 7 cm long, massive, pale orange with large round white patches, it has an opening bordered by big white teeth. Considered a living fossil, it has invaluable.

Figure 67: *Zoila (Thersites) contraria*, a rare cowry from Western Australia, much sought after by collectors.

Today, thanks to trawling activities in the depths of the Philippines seas, they are collected more frequently. Even the Pacific *Lyncina aurantium* was once considered very rare, living in deep waters outside the reefs. Between 1850 and 1880 300-400 dollars were paid at auction. James Cook was the first to bring this shell in Europe. In 1970, approximately 300 specimens were known (Saul, 1974), but diving has provided more specimens, especially in Melanesia and Philippines and the species is now considered quite common. However, it remains among the most sought after for its beauty, if not for its rarity. You can now buy it for 30-60 €, depending on its size and beauty. Being very much in demand, these quotes are higher than those of shells of equal rarity that are priced at 10-20 €.

Another "historic" cowry is *Leporicypraea valentia,* of which very few specimens exist. It was long known only by a solitary specimen at the British Museum and labelled "The Brindle Cowry of the Persian Gulf". Its origin was not certain for a long time, until it was found in the Torres Strait, New Guinea. It has a characteristic yellow-pink colour that becomes deep gold in the centre with a spectacular blood-red stain. Its malacological story is full of anecdotes in the mid-19th century: 50 gold pounds were paid for it in 1850, and 75 in 1917.

The story of one specimen, now in the *Musée Royal d'Histoire Naturelle de Belgique* in Brussels, is curious. In Australia, a *Leporicypraea valentia* specimen came into the possession of a certain Mr Hargraves who, not understanding its value, gave it to his children to play with: it is obvious what end it came to. A second specimen was offered, to a well-known collector from Sydney, a James Charles Cox (1833-1912), who bought immediately it. In 1905, he decided to sell his collection at auction in London and *L. valentia* was bought by the famous Belgian malacologist Philippe Dautzenberg (1849-1935), and for this reason, today this shell belongs to the collections of the Museum of Brussels. This shell, although not so rare, still retains a record: it is the species for which the highest price was paid for a shell. In 162 years, since 1811 when the first specimen was found to 1973, eight specimens existed. When in 1974 the ninth was found, it appeared at a very respectable price, more than 8,000 dollars (Raybaudi, 1979). Only five years after, as a result of new discoveries, the price fell to $ 2,200, dropping gradually up to 300-400 euro currently needed for a good specimen. Other cowries much prized by the market are *Lyncina nivosa* from the Indian Ocean, *Talparia exusta* from Red Sea and Gulf of Aden, *Umbilia hesitata* a famous and prized Australian species, and the South African *Cypraeovula fuscodentata.* Other cowries from deep waters (*Barycypraea fultoni, Erronea barclayi, Nesiocypraea teramachii*) are collected each year but with very few specimens, and are much sought after.

The cones

The family Conidae currently includes approximately 350-400 species, often of great beauty for their complexity of colours and design, prized by collectors, and holding their own with cowries and other highly-priced shells. Cones were the subject of intense collection on the part of indigenous peoples, especially the Indo-Pacific, when they realized that the "white people" were willing to pay for them. Whereas some cones can inject a poison, sometimes even dangerous to humans, poisoning cases are not common, but they must be handled very carefully. Some cones are not only beautiful, but rare. For collectors from all over the world, the shell of *Conus gloriamaris* has a very special meaning: it has been, for two centuries, the most sought-after and the rarest shell in the world. It first appeared at auction in 1757, where it was commercially called the "Glory of the Sea" (Fig. **68**). The beauty of the design and its rarity immediately made it the object of desire of all European collectors: in 1792, a fanatic Danish collector, Christian Hees Hwass (1731-1803), bought a *C. gloriamaris* at an auction only for to destroy it, so that the specimen that he already possessed should be still rarer. The beautiful and lucky name was given by the German malacologist Johann Hieronymus Chemnitz (1730-1778) when, twenty years after its appearance in Europe, he described it scientifically in 1777.

In 1837 Hugh Cuming (1791-1865), the great English collector, exclaimed: *I almost fainted with delight!,* when he collected three specimens living under a stone in the reef of Bohol in the southern Philippines. Unfortunately, in a short time, an earthquake shook that reef into the sea and this catastrophic event, of course, favoured the myth surrounding this shell. Only twenty specimens were known until the 1950s and one of these, never recovered, was stolen in 1951 during an exhibition at the American Museum of Natural History in New York. In those years, some specimens were also collected in New Guinea and, in September 1969, nearly one hundred were found at Guadalcanal in the Solomon Islands.

Figure 68: Two famous shells. *Conus gloriamaris* (left). *Epitonium lamellosum*, a wentletrap sea snail (right).

The rarity of *Conus gloriamaris* is now a past story: today you can buy a beautiful specimen for 40-50 Euros, but it still remains one of the most famous shells of the world. Another *Conus* in the history of malacology is *Conus cedonulli*, known by collectors as the "matchless cone", it was highly prized in the 1700s, and always sought after by collectors: in reality, it is only a set of chromatic forms of several species, including *Conus domenicanus*, and now its value is minimal. Today, *Conus milneedwardsi*, the "Glory of India", is the most prized cone. A specimen collected in 1903 in the waters of Bombay, is now housed in the National Museum of Wales; another was picked up in the 1930s from Mauritius. Today it can be found, always rarely, in the nets of trawlers operating off the coast of Mozambique. Beautiful specimens may appear on the international market, priced more than 200 Euro.

Figure 69: *Epitonium scalare*. The ribs of the sculpture make the shells of the epitoniids (the wentletraps) very stylish and sought after by collectors.

The muricids

Muricids belong to another large family that includes, worldwide, approximately 1,300 species (Houart, 2001). Their sizes vary considerably, the largest can exceed 30 cm, and their morphology is very complex. They can be globular or tapered, sometimes smooth, but more often their shell is bristling with thorns or winged expansions. The most rare and sought-after species are *Pterynotus loebbeckei*, from south-east Asia and of which only a few specimens are known, and *Chicoreus mergus*, *C. spectrum* and *Poirieria nuttingi* from the Caribbean and tropical Atlantic. A well-known muricid, highly sought after by collectors around the world, is *Murex pecten*, also 10 cm long, characterized by long and slender whorls that make it truly spectacular. It is distributed from the Red Sea to the Sea of Japan.

The volutes

Approximately 200 species of volutes are known and amateurs are often enthusiastic collectors of these shells, including some rare and valuable species. At least twenty are very rare, partly because they often live in deep water and are randomly collected by trawling; among the rarest afe *Cymbiolacca perplicata* from the Coral Sea, Australia's *Volutoconus bednalli* and the Indonesian *Cymbiola cymbiola*. They are named for the folds upon the columella, a character that they do not monopolize by any means. The spire is always prominent and has a rounded apex. They are handsome, aristocratic-looking shells, of graceful form, good size and elaborate ornamentation. The Imperial Volute (*V. imperialis*) wears a crown of upturned hollow tubercles on the outer three whorls of the peaked spire. This shell has always received the admiration of conchologists, having different nicknames such as “The Chinese Emperor’s Crown or The Crown of the Great Mogul” (Rogers, 1908).

The Courtier Volute (*V. aulica*) has the elegant shape of mitres, tapering gradually to each extremity. For a long time, this species was known only by a single specimen in the Duchess of Portland’s famous collection, which is now in the British Museum. Then Mr. Cuming obtained some beautiful specimens from the Sulu Archipelago. In the mid-19th century, the “Junonia Volute” (*V. junonia*) was considered rare, confined to deep waters, and found nowhere but off the east and west coasts of Florida. The possessor of a perfect specimen with its spots still dark and bright is to be congratulated. Once the demand for these shells was so great and the supply so short that a perfect specimen of good size would sell for $200 at the beginning of the 20th century (Rogers, 1908). “The Music Volute” (*V. musica*) has its whorls adorned with sets of parallel spiral lines, which look like bars of music, set the ‘ordinary’ distance apart. The typical form has its "notes" grouped in single lines, directly above and below the bars. Fine dots are thickly scattered between the two rows of large spots. The ground colour is of a creamy flesh tint.

The thorny oysters

Bivalves cannot reach, from a collecting point of view, the interest that gastropods have. In general, in fact, their shell is less spectacular and often dull. Exceptions are the thorny oysters, a family of bivalves with approximately 50 species, found mainly in tropical waters, often characterized by long spines. They are considered the king of the bivalves. Among the most important collection thorny oysters is *Spondylus regius*, a species which caused a small family drama. In the 18th century, it was considered a very rare species, and the few specimens that arrived in Europe from the Philippines and Japan were fought over by leading collectors for amazing prices. The French botanist Louis Claude Marie Richard (1754-1821) was an avid malacologist and possessed many very rare horny oysters.

Figure 70: *Spondylus regius* depicted by Jean-Charles Chenu in his *Manuel de Conchyliologie et de Paléontologie conchyliologée.*

When the opportunity arose to buy the splendid specimen of *Spondylus regius* depicted in the famous *Encyclopédie raisonné des sciences, des arts et des métiers* by Diderot and D'Alembert (1751), for 6,000 francs he did not hesitate to sell the family silver to possess it. Figuier (1882) said that when he was faced with his infuriated wife in a moment of distraction, he sat on the parcel containing the precious shell, breaking two spines. Professor Richard almost had a heart attack, and his wife ended up forgiving him. In 1819, Richard sold the shell to Victor Massena, son of the Duke of Rivoli, the famous General of Napoleon. Later, this shell, now of historical value and interest, belonged to Jean-Charles Chenu (1808-1879), the French author of the *Manuel de Conchyliologie et de Paléontologie conchyliologée* (1859-62) (Fig. **70**). Today, the specimen is in the collections of the Natural History Museum in Geneva.

This species is no longer one of the great malacological rarities, but *Spondylus americanus* living from Florida to Brazil is, as well as the rare *S. princeps,* living from California to Panama and *S. sinensis*, from the Indo-Pacific.

The scallops

A second family of bivalves valued by collectors is that of Pectinidae, to which the scallops (*Pecten* and allied genera) and *Chlamys* belong. Their rounded shell is characterized by radial ribs that give it a distinctive appearance of a comb. Many scallops show an orange colour and are spotted with stains of various colours. They can reach a size of 8-10 cm. In particular *Mimachlamys sanguinea, Mirapecten mirificus,* and *M. moluccensis* from the Philippines, *Manupecten pesfelis* from the Canary Islands, and the central Atlantic *Nodipecten nodus gabonensis* are very popular among collectors. Many of these species are of considerable commercial interest throughout the world.

Figure 71: A card showing the blue mussel, one of the most farmed bivalves in the world.

CHAPTER 10

Food for Man: Cooked and Raw

Abstract: At the dawn of civilization, Man was an active consumer of bivalve. Fishing is in fact one of the most ancient human activities and it is easy to imagine that the Palaeolithic man, arriving to the coast, quickly learned to recognize and appreciate limpets, mussels, clams, oysters, and other edible shellfish, easily to find along the shore.

The numerous shells, and not just marine, which are found today in heaps of waste (kitchen middens), leftovers of prehistoric dinners, emphasize the importance of shellfish as a food source for humans.

Virtually, all shellfish are edible and for many centuries were a tasty source of cheap food.

The oysters are certainly the "queen of shellfish": their taste, delicate or strong, varies depending on the different species, the area and quality of the waters. Also mussels are great "sea food": more humble than oysters, are very tasty. Moreover, scallops are much appreciated all over the world, and represent one of the main food products everywhere, especially in cold temperate zones. In addition, in many parts of the world, several bivalves, such as clams, cockles, razor clams and donacids are sought after and appreciated.

Some gastropods are very popular in the kitchen of different people, and abalones, muricids, and winkles are very much appreciated. Cuttlefish and squid are also important sources of food in some regions and could become one of the main sources of protein in the coming decades.

Keywords: Abalones, belon du belon, clams, clam-bake, cockles, *Crassostrea,* cuttlefish, donacids, *embractum baianum,* gastronomy, geoduck, gros blanc of Burgundy, *Helix,* Japanese oyster, kitchen middens, *Mercenaria,*molluscs, muricids, oyster, *Ostrea,* quahog, *Perna*, Portuguese oyster, razor clams, shell heaps, snail, squid, *Tapes*.

Riccardo Cattaneo-Vietti, Egidio Trainito and Mauro Doneddu

"It's a very remarkable circumstance, Sir," said Sam, "that poverty and oysters always seem to go together." "I don't understand you, Sam," said Mr. Pickwick. "What I mean, Sir," said Sam, "is, that the poorer a place is, the greater call there seems to be for oysters. Look here, Sir; here's a oyster-stall to every half-dozen houses. The street's lined with'em. Blessed if I don't think that when a man's very poor, he rushes out of his lodgings, and eats oysters in regular desperation".

Charles Dickens, *The Posthumous Papers of the Pickwick Club* (1837)

Cosimo said: - I said I do not want and do not want to! - and pushed away the plate of snails. Never such a serious disobedience had ever been seen.

Italo Calvino, *The Baron in the Trees* (1957)

Apart from the Baron Cosimo Piovasco di Rondò who had his good reasons for refusing to eat the snails cooked by his sister Battista, molluscs have always been very popular on the table: some reports suggest that they already formed part of our diet from approximately 300,000 years ago (Colones *et al.,* 2010). At the dawn of civilization, Man was, in fact, an active collector and consumer of molluscs, mainly bivalves, and since then, nothing has changed: the shellfish are still a major source of protein for many coastal populations, as well as the product of great gastronomic interest. Fishing is in fact one of the most ancient human activities and it is easy to imagine that the Palaeolithic man, arriving to the coast, quickly learned to recognize and appreciate limpets, mussels, clams, oysters, and other edible shellfish, easily found among the rocks, or in the sandy bottoms of shallow waters, accessible at low tide (Fig. **72**).

Figure 72: A romantic view of the English seashore during the low tide showing the richness of molluscs painted by Eileen Soper (1905-1990), the popular English etcher and illustrator of children's and wildlife books.

Probably, in some particularly favourable environments, such as bays or along the banks of lagoons and coastal ponds, the wide availability of shellfish alone would have convinced man to become more sedentary, abandoning the hunting that required wandering habits (Fig. **74**).

Figure 73: Bocho Island (Malaysia), the most important industry is squid fishing.

Figure 74: Collecting clams at low tide in Galicia (Spain).

Shell heaps

The numerous shells, and not just marine, which are found today in heaps of waste, leftovers of prehistoric dinners, emphasize the importance of shellfish as a food source for humans. These deposits, known as “kitchen middens” by archaeologists, indicate unequivocally the presence of the most ancient human settlements along the coast. When the shells become predominant in these deposits, they became known as “shell heaps”. In Niath Great Cave of Sarawak, Malaysian Borneo, these deposits have been dated to approximately 40,000 years ago (Saul, 1974), while in Japan, in the regions of Kanto and Hokkaido, the same, called “kaizuka” were dated from 17,000 to 4,000 BC. In Europe, the oldest deposits date back to the late Mesolithic period, 7,000 years ago, and have spread everywhere, both in the Mediterranean and along the Atlantic coast of Europe, from Portugal to Scandinavia. Shell heaps of freshwater bivalves, the unionids, are also common along the Scottish rivers, while shells of land snails are common in caves inhabited in prehistoric times in Denmark, Brazil, Tierra del Fuego, Australia, and New Zealand. In America, these deposits are still visible along both the Atlantic and Pacific coasts: those of New England are usually made by the "quahog," *Mercenaria mercenaria*, the North America clam, one of the main foods of the local Indian tribes: in some cases, these shell heaps can reach a length of 50 m, a width of 20 m and height of 5 m, veritable mountains of shells (Ristine, 2009). These great deposits, dating back more than 2,500 years, can be seen in Damariscotta, Maine (USA), an area once inhabited by the Abenaki ethnic group to whom these great shell heaps were attributed. Other massive deposits can still be seen in St Petersburg in Florida. These shell heaps were so numerous that, at the beginning of the 20th century, they were used as road surfaces, thus destroying an important archaeological heritage. Between the end of the 19th and the beginning of the 20th century, the consumption of clams, very popular among the local ethnic groups, became popular also among the Anglo-Saxons of the eastern coasts of North America. Consequently, clamming along the Atlantic beaches become the favourite pastime during week-ends and the "clam-bake", the cooking of clams and many other marine organisms in large holes in the sand, an Indian custom adopted by the Pilgrims upon their arrival in Massachusetts, become a real moment of relaxation: discarded shells were left on the beach. Also along the western United States and particularly in San Francisco Bay and Puget Sound, in Washington, large deposits of Olympia oyster shells, *Ostrea lurida*, abundantly consumed by the Chumash people for thousands of years, are still present.

The remains of these large banquets continued, inevitably, to accumulate on the beaches and today, if the consumption of shellfish is great and the service of garbage modest, in the vicinity of restaurants, empty shells accumulate (Fig. **75**).

Figure 75: A shell heap on a beach in front of a restaurant at La Serena (Chile).

Shell heaps have considerable scientific interest not only as tell us about the customs and traditions of the past, but they also provide information on the dynamics of natural populations of species eaten, subject to harvesting for very long periods. The study of the *Strombus gigas* deposits at Los Roques Archipelago National Park in Venezuela (Schapira *et al*., 2009) has, for example, demonstrated that from the pre-Columbian age to the present, the size of these large gastropods has gradually decreased and now, the survival of these populations is at risk.

A tasty source of protein

Virtually all shellfish are edible and for many centuries, oysters and mussels (Fig. **71**), so common along the coast and at river mouths, were a tasty source of cheap food, rich in protein and a high content of zinc, magnesium, iodine, and iron. They show organoleptic characteristics very different from each other and difficult to describe: the iodized salty taste of the oyster, the softness of a "abalone", the delicacy of the scallops, the fleshiness of the octopus, the toughness of the muricids: large differences in flavours that make each species unique. Some shellfish are eaten raw (Fig. **76**), seasoned with a bit of lemon, others are boiled, steamed or roasted, or end up in soups and sauces. Sometimes, molluscs are also preserved in brine or dried, and especially for the Eastern markets, frozen mussels are very popular. In Roman times, they were generally not eaten raw: *embractum baianum*, according to Pliny the Elder, was a kind of stew of seafood containing oysters, thorny oysters, and the common sea anemone "hair of Venus" (*Anemonia sulcata*). It was a typical recipe of the city of Baia, near Pozzuoli (Naples), already home to important fish and oyster farms in the Imperial Age (Battiato, 2012).

Some foods, so popular in some Asian populations, are a little bit strange, at least for western tastes. While for some shellfish, like oysters, clams and calamari, infact, their gastronomic importance is clear, some Oriental choices, at least for the Western mind, are very strange. The indigenous peoples of the Kurile Islands of the Aleutian Archipelago roast a big nudibranch, *Tochuina tetraquetra*, while the inhabitants of the Fiji Islands in the Pacific Ocean collect sea slugs, especially *Dolabella auricularia*, and their eggs, to boil in coconut milk. Even the dried squid which are sold at the fish market in Tokyo not would raise much interest in Europe. Another “strange” species is the impressive geoduck (*Panopea generosa*), a "monstrous" bivalve typical of muddy coasts of the northwest Pacific. They are extremely abundant in the inland waters of Puget Sound, British Columbia and Alaska, where the sub-tidal populations support important commercial fisheries. The clam's name is of Native American origin and means "dig deep". It can reach extraordinary dimensions (for a bivalve), weighing more than 7 kg: in the Far East, is considered a true delicacy. Its success is also due to the fact that the Chinese think that, thanks to its vaguely phallic shape, it has strong aphrodisiac properties, promoting sexual performance. Its cost reaches 60 €/kg and feeds a market of nearly 80 million dollars a year. Today, its harvesting is seriously regulated in United States: the exploitation of its beds was, in fact, too intense and the cycle of growth of this species to reach a size of commercial interest is quite long, more than 10 years. It can be fished only for certain annual quotas, in periods not during reproduction, and the specimens caught must have a minimum size. For several years, sea farming was also attempted, with low success due to the long span life of the species. They attain their maximum size by approximately 15 years, and can live at least as long as 170 years.

The “queen of shellfish” is certainly the oyster, much appreciated since the Palaeolithic Age: its taste, delicate or strong, varies depending on the area and the quality of the waters. The excellence of their meat and their capacity to form large beds along the coast and in shallow waters, which are easily accessible, determined their fishing interest since ancient times: with the pearl-oysters (which have nothing to do with the edible ones), they are the most important commercial molluscs.

Different species are consumed in all continents; among the most important are the European "flat" oyster, *Ostrea edulis*, the Japanese" or "Portuguese" *Crassostrea gigas*, the aastern American *C. virginica*, the Olympia oyster, *O. lurida,* from North American Pacific coasts, *O. chilensis* from the South Pacific, and *C. rhizophorœ,* typical of the tropical Atlantic mangroves which is replaced, in the Indo-Pacific, by *C. cucullata.*

In Europe, the most sought-after oyster is the endemic *Ostrea edulis*, with its characteristic round shape which is commercially known as the "flat" oyster.

It lives cemented to the substrate with its right valve in the Mediterranean and along the Atlantic coasts of Europe. Being a hermaphroditic species, a single specimen produces both female and male gametes, reaching a full maturity at a water temperature of approximately 16 ° C. The fertilized eggs are not immediately released: only when the larvae have reached a certain developmental stage will they leave the parent, beginning the second phase of their development, which is pelagic and lasting approximately two weeks. At the time of the metamorphosis, the larva sinks to the bottom and looks for a hard substrate on which to settle.

In the northern Adriatic Sea, peaks of settlement are during May and June. Its growth is relatively rapid in the first eighteen months, reaching nearly 50 grams of total weight. In recent decades, along the Adriatic coast, the natural beds were overfished, but encouraging results have been obtained in farming, with semi-floating plants.

At Arcachon (Atlantic French coasts), it is known as "gavette", while in Brittany it becomes the famous "belon" or the even more famous "belon du belon", a population of *Ostrea edulis* aged for months in the river of the same name. The "Japanese" farmed oysters have a more delicate flavour.

Figure 76: A raw seafood dish in an Italian restaurant.

Its farming has spread everywhere, but those growing in the Marennes-Oléron, French department of Charente-Maritime, the centre of Europe's most important farming, are protected by a Label Rouge, which guarantees their superior quality: the famous "pousses en Claires" and especially the “fines de Claires” or *huîtres vertes*. In this case, a “blue” diatom, *Haslea ostrearia*, present on the gills of the “Claires”, gives a characteristic green colour due to the concentration of a pigment called “marennine”. Thanks to this “affinage”, it is possible to improve the oyster quality, just before sale. Another variety very much appreciated by connoisseurs is the "fine Binic", farmed in open sea in the Bay of St. Brieuc, in northern Brittany: it is very salty and tasty. Then, there are the so-called "sauvage" oysters collected directly from natural beds: very expensive because of their rarity, and called "pied de cheval" when they exceed the pound in weight. In England the "Whitstable oysters”, from the Thames estuary, are particularly appreciated and also protected.

Generally oysters are consumed “natural”, raw, at best with a little of lemon juice: Tzar Nicholas the First (1796-1855) liked to sprinkle them generously with black pepper, also adding half a teaspoon of Royal Golden, the most valuable caviar from the Caspian Sea. On the contrary, in America oysters are often grilled, while in Apulia (Italy), until the 19th century, they were fried or preserved in sour-sweet sauce. Even in Brittany, the home of the oysters, they are often cooked and served with a sauce made with shallots and chives, washed down with white wine.

Oyster are available all around the year, but in Europe, the rule is that the best months to enjoy them are those with the "r" in the French language and therefore is not recommended to consume them in summer, from May to August, when they are full of "milk", which is when the gonads are ripe, rich in fertilized eggs.

Also mussels are great "sea food", which has been part of the human diet everywhere and in all times: shells of these species are, in fact, common in the heaps of waste from prehistoric kitchens. The Atlantic *Mytilus edulis*, the Mediterranean *M. galloprovincialis,* the various species of *Perna* from Southern Seas, are generally very tolerant to environmental changes and so may create naturally high dense populations within a few feet of water, especially in coastal lagoons, where, thanks to the availability of organic matter in suspension, they grow rapidly. The European mussels, of a beautiful orange colour in females, changing into a pale yellow in males, are more humble than oysters, but very tasty. In Belgium, mussel is a national dish, while the "moules de Bouchot de la Baie du Mont Saint Michel" in France, boast of an Appellation d'Origine Contrôlée.

In general, mussels are steamed, though they are often eaten raw especially in southern Italy. In the famous *impepata* of mussels, a classic Neapolitan recipe, the shellfish are opened on a fire with a base of olive oil and garlic, and sprinkled with freshly black pepper.

All around the world, mussels are now farmed and the fishing for "wild" mussels is limited, mainly due to the strong competition with farmed ones: however, *Mytilus edulis* remains among the 15 species fished in Europe, reaching a production of approximately 75,000 tons/year (source: Eurostat). In the Adriatic Sea, a professional fishing on natural mussel beds survives along the submerged reefs of the Conero Promontory (Ancona), where these bivalves, locally called "Portonovo wild mussels", are harvested from April to October. This kind of mussel is now a Slow Food Presidium and locally very appreciated. The so-called "bearded mussels" (*Modiolus barbatus*), with excellent qualities, is often eaten raw, but rare on the market, while the date mussel (*Lithophaga lithophaga*), whose soup is considered a real delicacy, is now protected: since 1988, the harvesting of this species has been rightly banned in Europe to safeguard the rocky coast where the date mussel lives.

In addition to oysters and mussels, in many parts of the world several bivalves, such as clams, cockles, razor clams, donacids (belonging to *Cardium, Solen, Tapes, Donax, Mactra, Mercenaria, Venerupis, Chamaelea* and allied genera) are regularly consumed, often served alone, sautéed or fried, or used for a sauce for pasta.

The warty Venus-clam, *Venus verrucosa,* cannot bear heat, and is never cooked. The smooth Venus-clam (*Callista chione*) may be eaten both cooked and raw (Fig. **77**). The Brazilian venerid "berbigão" (*Anomalocardia brasiliana*) is famous being a "Slow Food Presidium": it is the basis of traditional economics, supporting hundreds of thousands of families. It is harvested by fishermen with a rake with spaced teeth so as to select only the individuals of a certain size, according to the rules of good sustainable harvesting.

Figure 77: The Mediterranean smooth Venus-clam, *Callista chione*, may be eaten either cooked or raw.

Scallops (*Pecten* (Fig. **78**), *Argopecten, Placopecten, Patinopecten, Aequipecten, Mimachlamys*) are much appreciated all over the world, and represent one of the main food products everywhere, especially in cold temperate zones: they can be served raw, the adductor muscle, but more frequent are baked “au gratin” or sautéed. In Europe, of the Atlantic scallops, the "Coquilles St Jacques de Côtes-d'Armor" in Brittany are particularly appreciated and protected.

Figure 78: Scallops from Aquitaine (Atlantic France).

Other bivalves of commercial interest live on rocky coastlines, such as the Mediterranean thorny oyster (*Spondylus gaederopus*) which must be consumed raw, as well as the arks (*Arca noae, A. barbata*), which are much sought after, but their consumption is mostly localised. In particular, the thorny oysters are loved all over the world and not just today: in the ancient Ecuadorian culture, the *mullu*, *Spondylus calcifer* and *S. princeps* was called “the food of the Gods" according to a Quechua myth recorded in the 17th century.

Over all, the smooth Venus-clam, *Callista chione* (Fig. **77**), the cockles (*Cardium edule, C. tuberculatum*), wedge shells (*Tellina, Donax*), razor clams (*Solen vagina, Ensis ensis*) and especially the common clam (*Chamaelea gallina*) of which are collected annually at 30,000 tons in Italy and 50,000 tons in Turkish waters (source Eurostat). Since the seventies of the last century, the use of "deadly" fishing equipment, such as the hydraulic dredge or "turbochargers", has spread to the Adriatic Sea: it completely removes the sandy bottom to suck shellfish onto the surface. The obvious and serious damages caused by this kind of gear to the environment and to the same mollusc populations are unquestionable: in Europe, the rules governing fishing with "turbochargers" establish a minimum landing size of 25 mm, and a ban on fishing less than 0.3 miles from the coast (EC Reg. 1967/06). Furthermore, the dredgers must have a width of less than 3 m, a maximum weight of 600 kg and must be equipped with selective devices to facilitate the escape of younger clams (Bombace and Lucchetti, 2011). Finally, most of the Mediterranean bivalves of food interest live sunken in coastal sandy or sandy-muddy bottoms, and before being commercialized must be "purified" in appropriately centres (**Chapter 11**).

Gastropods on the table

Man consumes mainly bivalves and cephalopods: gastropods represent only 2% of the molluscs caught in the world (Leiva and Castilla, 2002). Some of these, however, are very popular: abalones (*Haliotis*) are very much appreciated (Fig. **80**), and their fleshy foot is cut into slices and fried in California, Mexico, Japan, South Africa, and Australia.

Even in Europe, both along the Atlantic coast and Mediterranean, the Mediterranean abalone (*Haliotis tuberculata*), known in Sicily as "bull's eye", is much appreciated, along with limpets (Fig. **79**). Smaller than its tropical cousins, it is less tasty. Unfortunately, it is often collected by poisoning the seawater with copper sulphate that pushes the molluscs to come out from under the stones where they hide during the day. This illegal fishing technique has damaged many populations along the Italian coasts and the species is certainly in trouble in many areas.

Even in France, the harvesting was excessive and today fishing is regulated: the fishermen pay a fixed annual personal fee and the catches cannot exceed a total of 14,000 units/year.

Figure 79: Mediterranean limpets in a Sicilian market.

Throughout the Mediterranean Sea, some muricids (*Hexaplex truncatus and Bolinus brandaris*) and some whelks are often eaten cooked. In the Adriatic Sea, a mud snail, *Nassarius mutabilis,* locally called "bombolino", is particularly appreciated as well as the "common pelican's-foot" (*Aporrhais* spp.), a gastropod disinguished by pointed processes on the outer lip of the shell, hence the common name.

The Atlantic periwinkle, *Littorina littorea* (Fig. **81**), a small gastropod common on the rocky coasts, is generally consumed in France, Belgium, and England. Its name, periwinkle, is supposed to be modified from "petty winkle," the small one, to distinguish it in the London markets from the large winkle, or whelk, *Buccinum undatum*. Both were staple foods in England: thousands of tons of this tiny little mollusc were collected each year on the rocky coasts of the British Isles, and sold from corner stalls or push carts in the streets. In Ireland, it is called "wilk" and it is not difficult to buy it, roasted and seasoned with salt, on the beach.

Figure 80: Fresh abalones in a Japanese market.

In the Caribbean and other tropical sites, millions of "Queen conch", *Strombus gigas,* were fished, especially for roasting, as well for the shell, used for decorative purposes. Today, restrictive laws favour the conservation of this species. Another gastropod highly sought after in Chile is a muricid, *Concholepas concholepas* (Fig. **109**), locally called "loco", also very appreciated on the Japanese market and whose populations have suffered a devastating overfishing in recent years.

Figure 81: The Atlantic periwinkle, *Littorina littorea*, is generally consumed in France, Belgium, and England.

Everywhere, land snails are collected manually, even if, in ancient times, the Romans had already developed their farming (**Chapter 11**). In the Mediterranean region, in addition to the well-known and widely farmed species of the genus *Helix* (the "petit gris" and the "gros blanc" of Burgundy), the dune snail, *Theba pisana*, is also collected and locally consumed. These species with a high demographic potential and an omnivorous behaviour, have been introduced in northern Europe and in almost all continents, where it is easy to find even in urban environments.

Cephalopods: a food for future?

Cephalopods are very valuable and should be almost always served cooked, albeit briefly. Only the octopus in the Mediterranean is often eaten raw. The initial preparation of the octopus is crucial: it must be beaten with force to pulp and soften the meat, or frozen and thawed several times. The "Galician octopus" is a famous recipe in Spain, boiled and flavoured with paprika or chilli. In Naples, the "Luciana style octopus" is cooked in its water and takes its name from an old seaside district.

The cuttlefish and the squid are also important sources of food in some regions (Figs. **73**, **82**). They are often fried, or served stuffed or baked. Squid ink must not be forgotten, often used in the famed Italian "risotto": some cooks even knead the ink with flour to make a kind of black bread.

Globally, cephalopod fishing, and in particular squids, has had an extraordinary development in recent decades, becoming an important source of income for many countries: in 2002, more than 2 million tons of cephalopods were caught in the world (FAO, 2005). Their availability in the ocean is such that it is likely to be one of the main sources of protein in the coming decades: in fact, the decrease of their large predators, namely sharks, pelagic fish, and cetaceans, has certainly favoured the quantitative development of these populations which show a rapid reproduction rate and significant growth velocity (Caddy and Rodhouse, 1998).

Two species, *Illex argentinus* (the south-western Atlantic squid) and *Todarodes pacificus* (the Japanese flying squid) represent approximately 45% of world catches and show huge fluctuations in the annual catch not always predictable.

Another important commercial squid is *Loligo gahi*, the Patagonian squid, and common to both the Atlantic Ocean and the South Pacific, whose harvesting is approximately 50,000 tons per year.

Figure 82: Dried squids in a Japanese market.

In the Mediterranean, and in particular in the Adriatic Sea, the fishing of cuttlefish (*Sepia officinalis*) is of particular interest: in spite of the decline in production in recent years, it is still a significant part of the landing: it accounts for nearly 20% of the annual catch of the Italian small fisheries, especially during the winter season.

The main marine mollusc (gastropod and bivalve) global production in the wild or farmed (source FAO, data modified).

Scientific name	Common name	Production (t/y)	Production zone
Crassostrea gigas	Japanese oyster	4.400.000	From Japan, farmed everywhere
Ruditapes philippinarum	Philippines clam	3.000.000	From Philippines, farmed everywhere
Patinopecten yessoensis	Yesso scallop	1.400.000	Asia
Mytilus planulatus	Australian mussel	765.000	Australia, Oceania
Pecten maximus	Atlantic scallop	648.000	Europe
Mytilus edulis	Blue mussel	500.000	Europe
Mactra veneriformis	Surf clam	429.000	Asia
Crassostrea rhizophorae	Mangrove oyster	417.000	Atlantic Tropical America
Anadara granosa	Blood cockle	400.000	Asia
Ostrea chilensis	Chilean oyster	167.000	South America
Mytilus galloprovincialis	Mediterranean mussel	160.000	Europe
Ostrea lurida	Olympia oyster	136.000	Pacific North America
Aulacomya ater	Cholga mussel	120.000	South America
Pecten novaezelandiae	New Zealand scallop	120.000	New Zealand
Choromytilus chorus	Choro mussel	117.000	South America
Pecten fumatus	Australian scallops	110.000	Australia
Protothaca staminea	Pacific littleneck clam	98.000	North America
Crassostrea virginica	Eastern or Virginia oyster	90.000	Atlantic North America
Perna viridis	Green mussel	90.000	Asia
Meretrix lusoria	Oriental common clam	77.000	Far East
Perna perna	South American green mussel	75.000	South America
Perna canaliculus	New Zealand green mussel	70.000	New Zealand
Argopecten ventricosus	Pacific scallop	41.000	North America
Placopecten magellanicus	American scallop	40.000	North America
Mytilus coruscus	Korean mussel	40.000	Asia
Aequipecten opercularis	Queen scallop	38.000	Europe
Mya arenaria	Sand gaper	32.000	North America
Saxidomus giganteus	Butter clam	14.000	North America
Saccostrea cuccullata	Hooded oyster	12.000	Africa
Crassostrea iredalei	Slipper cupped oyster	12.000	Asia
Argopecten purpuratus	Peruvian scallop	11.000	South America
Corbicula japonica	Japanese basket clam	11.000	Asia
Mytilus chilensis	Chilean mussel	10.000	South America
Mercenaria mercenaria	Northern quahog	8.000	Atlantic North America
Tapes decussatus	Grooved carpet shell	7.000	Mediterranean Sea
Saccostrea commercialis	Sydney rock oyster	6.000	Australia
Haliotis tuberculata	European abalone	6.000	Europe
Ostrea edulis	Flat European oyster	6.000	North America, Europe
Haliotis midae	Abalone perlemoen	5.000	South Africa
Cerastoderma edule	Mediterranean cockle	2.000	Mediterranean Sea
Venerupis corrugata	Atlantic clam	2.000	Atlantic, Europe

CHAPTER 11

Farming Molluscs

Abstract: The bivalve farming started in Europe during the Roman period and probably before in Asia.
At the end of the 19th century, the consumption of molluscs became very popular everywhere, while the use of motor boats, with which fishermen could dredge more easily and deeper, quickly destroyed the natural beds of oysters and mussels on both sides of the Atlantic. Moreover, the pollution of water, especially in estuarine areas, following the industrial revolution, also contributed to their destruction. It was clear that if people wanted to continue to consume large quantities of bivalves, it was necessary to begin to farm them.
Today, if we exclude a small number of land snails, only marine bivalves (mainly oysters, clams, scallops, and mussels) are farmed with great success, utilising methods sometimes old of centuries. Typically, shellfish farming is considered "environmentally friendly" as molluscs are not artificially feed, and consequently no surplus of organic substances is added to the environment.
Since 8,000 years ago, some land snails have been an important source of protein for humans and their roasted shells are often abundant throughout the Mediterranean region. During the Republican period, Romans farmed them with great success. At the beginning of the 19th century, land snails became a "must" in the French cuisine (and therefore in the world), and today some species of snails are farmed in different countries both in Europe (Italy, France, and Spain) and in tropical countries.

Keywords: *Achatina,* Big Oyster, Burgundy snail, *Choromytilus*, clam, Coste, *Crassostrea*, farming, *Mercenaria, Helix, Mytilus,* molluscs, *Patinopecten*, *Perna,* Philippine clam, *Pila, Ruditapes,* Yesso scallop, triploid oyster, *Venerupis, Tapes*, white snail.

Riccardo Cattaneo-Vietti, Egidio Trainito and Mauro Doneddu

I always thank the Creator that, with his mighty will, he created the oysters at sea and the Rhine dry white wine inland!

Heinrich Heine (1797-1856)

In the 19th century, the consumption of molluscs was very popular (**Chapter 10**), but the extraordinary growth of the European and North American cities along the coasts, as a result of the industrial revolution, and the accessibility to the oysters and mussels natural beds along the Atlantic coasts, provoked an uncontrolled exploitation and consequently, a rapid destruction of these habitats. In a short time, oysters became more and more rare and their price began to rise: from a popular and cheap food, they soon became exclusive and much sought after. In North America, the beds collapsed, not just those of oysters: populations of some very interesting commercial bivalves (*Mercenaria mercenaria, Spisula solidissima, Mya arenaria, Ensis directus*), present in easy-to-rake over muddy and sandy bottoms along the coast, suffered quickly from the continuous sampling. On both sides of the Atlantic, it was clear that if people wanted to continue to consume large quantities of bivalves it was necessary to begin to farm them, implementing forms of collecting and housing, and to maintain a regular flow of product to the market. Gradually, a number of bivalves began to be farmed, but this seafood would never been cheap.

The suitable species

Not all shellfish are suitable for farming. In addition to their commercial value, which must cover costs and leave a profit margin, a species must have different characteristics to be successfully farmed. Firstly, the species must not have a very complex life cycle: for example, the unionids, freshwater bivalves, large producers of mother-of pearl (**Chapter 7**), would potentially be species of commercial interest, but their larva, known as *glochidium*, must necessarily parasite a fish to complete its life cycle, so this certainly does not facilitate their farming. However, if the species is economically important, man has attempted to farm it. For example, the case of *Sinanodonta woodiana*, a native unionid from south-east Asia, accidentally introduced in the early sixties of the 20th century in Europe, is now farmed to produce freshwater pearls and mother-of-pearl with interesting results. Other experiments are being conducted in the United States, Japan, and China with other unionids, such as *Amblema plicata*, *Hyriopsis schlegeli*, *H. cumini* and *Cristaria plicata.* The most suitable species for farming are those that better withstand possible environmental changes, even sudden in occurrence, of chemical and physical variables of the water column and of sediments (e.g. temperature, salinity, turbidity, dissolved oxygen and organic matter levels) and which can support strong tides and exposure to air. Moreover, the candidate must also have a good growth rate, high reproductive capacity, good resistance to pests, and low mortality. Often the species living in brackish waters and lagoons, enduring significant changes in salinity, are the most "robust" and therefore the most suitable. The success depends, therefore, on a series of chemical, physical, and biological agents that must be appropriate to the needs of the species that you want to farm. Of course, in addition to a good water quality, the seabed features are important too: the grain size of the substrate, the concentration of organic matter in the sediment, and its rate of oxidation can be crucial to success. In addition, the bivalves are active filter-feeders, and able to accumulate pathogenic microorganisms, toxic chemicals, and heavy metals in their body, sometimes in concentrations sufficient to be dangerous to humans: their consumption, when farmed in contaminated environments, may be really hazardous (**Chapter 13**). The selection of the plant site and the control of water quality are, therefore, of great importance and must also take into account, so that it is not possible to realize plants in areas which may interfere with local fishing or commercial navigation. In other words, the choice of a site for a shellfish farm is rather complex.

Figure 83: The Japanese Yesso scallop, *Patinopecten yessoensis*, the most important farmed scallop in the world.

Today, if we exclude a small number of land snails, only marine bivalves are farmed, although not many species (Spencer, 2002), but some with great success. In 2006, the world production was estimated to be more then 14 million tons, of which approximately 60-70% from aquaculture. The production continues to rise, with an annual growth rate of nearly 4%, representing 27% of world aquaculture, at a turnover of 11 billion dollars. Globally, the shellfish farmed with most success are several species of oysters, but mainly *Crassostrea gigas*, with more than 4.4 million tons/year. This production is largely localized in China, while mussels (*Mytilus*) are cultivated in Europe, Australia and South America, and *Choromytilus, Perna* and even in South America and New Zealand) with 700,000 tons/year. Another species of great commercial interest is the Japanese Yesso scallop (*Patinopecten yessoensis*), with a production of more than 1.3 million tons/year (Fig. **83**), while the farming of the Atlantic scallop (*Pecten maximus*) is still in experimental stages, because some problems in the conditioning the new settled juveniles are not yet solved, so today they are harvested directly from nature, using traps and ranching them in controlled environments. Even the grainy arcid *Anadara* is widely farmed (over 400,000 tons/year), especially along the western coast of the Malay Peninsula and Thailand.

Typically, shellfish farming is considered "environmentally friendly" as man does not artificially feed the molluscs, and consequently does not add a surplus of organic substances to the environment as occurs, for example, in fish farming. However, the intensive systems have a significant impact on the landscape and produce a strong enrichment of organic matter in the sediments. A final problem certainly not to be underestimated is the introduction of non-native species, which could cause serious environmental problems (**Chapter 12**).

The farming of gastropods is rare, if we exclude the tropical abalones (*Haliotis midae, H. rubra, H. laevigata,* and *H. scalaris*), which began in the early 1960s in South Africa, China, Korea, Japan and Australia. However, their natural growth rate is slow and over-fishing and poaching, reducing the wild populations, favoured the introduction of farming techniques. Hence, the interest from investors is modest, even if the market value of abalones is very high: world production is approximately 7,000 tons/year.

The Roman oyster

The Chinese, a few centuries before Christ, began to farm mussels and oysters (Stott, 2004), favouring the settlement of their larvae on bamboo poles driven into the coastal strip, to the limits of the low tide. In Europe, the Romans started to house and perhaps farm fish and shellfish: designer and supporter of these activities was the Senator Caius Sergius Orata (140-90 BC). Pliny the Elder (23?-79 AD), in his *Naturalis Historia*, wrote that the first oyster farm was realized by Sergius Orata in Baiano, during the consulate of Lucius Licinius Crassus (140-91 BC), before the Marsican war: *Ostrearum vivaria primus omnium Sergius Orata invenit in Baiano, aetate L. Crassi oratoris, ante Marsicum bellum ....*

In fact, Senator Orata farmed oysters in Lucrinus Lake and probably also in Fusaro Lake, close Naples: the seed (the newly settled post-larvae) was collected using shards of pottery and the growth of juveniles was assisted by removing parasites and predators. According to Lucius Junius Moderato Columella, a Spanish expert on agricultural matters from the 1st century AD, Orata and his friend Licinius Murena had their own names from fish; the term "orata" in Latin means gilthead sea bream, while "murena" is the moray eel.

There is a story about Senator Orata that is worth recounting. Accused of stealing a quantity of oysters from a rival, Orata defended himself thus: *Yes, gentlemen of the court, Orata is guilty, but this one is, not me* (showing a sea bream, well-known predator of oysters, approximately 60 cm long with an oyster in its mouth) (Battiato, 2012a).

In the Imperial Age, farmed oysters became fashionable even at the Palace: Juvenal satirized the Emperor Nero, a famous consumer of oysters, that he was able to understand at the first bite the origin of an oyster, whether it had been brought up in the Circeo, Lucrinus Lake or came from far-away Rutupia, today Richborough, in Kent, on the east coast of England! *Circe nata forint, out Lucrino ad sauri, Rutupinoae editae fundo ostrea, callebat primo deprendere morsa.*

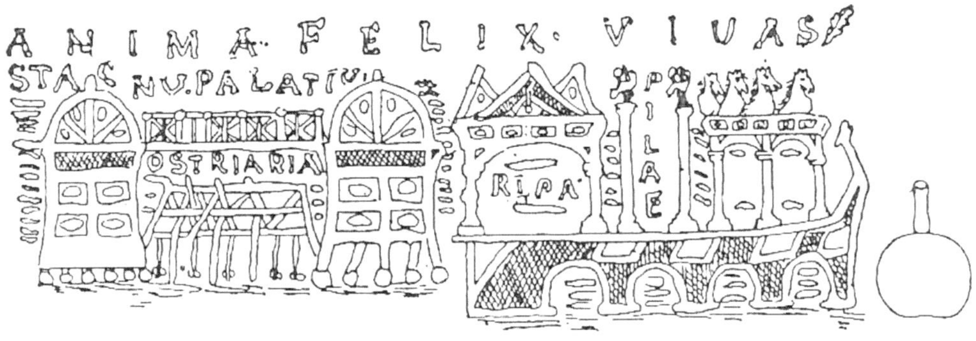

Figure 84: Engraving on a glass flask showing an *ostrearia* (oyster farm) of the Imperial Age in the Phlegrean Lakes near Naples.

An amazing testimony to the presence of oyster farming in the Imperial Age in the Phlegrean Lakes near Naples, called the "ostrearia" in Latin, is found in the engravings on some glass flasks (Fig. **84**), now preserved at the Museum of Glass in Corning (New York). After the conquest of Britain, where the consumption of oysters was already widespread, the Romans began to import them from the coasts of Kent by mule (Stott, 2004) along a "route de l'huître" crossing Gaul and punctuated by numerous nurseries, fed with sea water to store the oysters on their journey to Rome.

Oyster farming

At the end of the 18th century, along the Atlantic coasts of Europe, from Spain to Denmark and southern England, large natural oyster beds (*Ostrea edulis*), extending for miles, were a cheap food source for the local populations, just as it did on the shores of the North America, where *Crassostrea virginica* beds were ubiquitous. In the years immediately after the Civil War, America's cities grew at a furious rate in part because of a shift from agrarian to industrial economies, but also because of massive waves of European immigration. New York, for example, from 1890 to 1915 grew from 2.5 million to 6 million people, while the construction of the railway networks, through which they could easily reach the markets of the big cities, promoted trade. With the growth of populations, the demand for oysters increased, being a cheap tasty food, also considered aphrodisiac! Even in the 1980s, the 3,000 acre beds in Chesapeake Bay produced more than 25 million oysters each year (Saul, 1974), and oyster and clam bars arose everywhere in the United States: in Boston, Union Street, there is still the Union Oyster House, open since 1826 and now considered a national monument (Fig. **85**). At the beginning of the 20th century, New Yorkers consumed approximately 500,000 barrels of oysters per year, an average of 660 oysters per person per year (Restine, 2009). This is a really outstanding average, considering the actual request.

Before the 20th century, when people thought of New York, they thought of oysters: Mark Kurlansky wrote in his *The Big Oyster* (2006) that the "Big Apple" had been, in reality, the "Big Oyster", emphasising the importance that this mollusc played in the development of the city. The same occurred along the European side of the Atlantic: but slowly, all around the world, overfishing began to bear fruit: the annual harvest of oysters on the natural beds of Cancale, near Saint-Malo in Brittany, fell by 70 million oysters/year in the 1940s to 1 million in the 1960s, even if the Brest banks still produced 3 million oysters/ year before the first world war.

Throughout the 19th century, fishing on the natural oyster beds had taken place with simple gear, often carried out by individuals using simple collection techniques. Jimmy Morrison, a fisherman from Falmouth in Cornwall, said *dredging for oysters under sail in the Carrick Roads survives as one of the few things a man can do on his own, with his own boat* (Bird, 2010). The introduction of motor boats, with which they could dredge more easily, quickly destroyed the beds. Moreover, the pollution of water, especially in estuarine areas, following the industrial revolution also contributed to their destruction. In general, the situation got worse along both sides of the Atlantic: it is appropriate to recall that in 1905, a Tréguier in Brittany dredging on a particular oyster bank was allowed only one day per year. In practice, all vessels, on a particular day, could trawl for just 45 minutes, and everything was then suspended until the following year (Bouchet *et al.*, 1979).

Figure 85: The Union Oyster House sign in Boston, the oldest continually operated restaurant and oyster bar in the United States, open since 1826.

The reckless harvesting led to a rapid collapse of the natural production and the need to find new ways of obtaining the oysters, i.e. farming. In today's world, the main cultivated oysters are the "Japanese" or "Portuguese" *Crassostrea gigas,* and the American eastern oyster, *C. virginica.* The major production areas are now located in Normandy, Brittany, and Arcachon (Aquitaine), while in the French Mediterranean, important facilities are in the Thau Lagoon and Camargue. Minor productions, but of great prestige, are active in the Danish Limfjord, in Galway Bay in Ireland, and in some Dutch localities, such as Oosterschelde and Grevelingen. The annual French production now exceeds 500,000 tons with a business turnover of 600 million Euros. In Italy, the production is less than 500 tons per year: not enough to meet the demand of the domestic market estimated to be at least 5-10,000 tons per year and largely covered by imports from France, Spain and Netherlands. The main production centre is Taranto, where oyster farming was already active in the Roman period; after a long period of crisis, it reasserted itself in the 18th century and then disappeared at the end of the 1960s as a result of the industrialization of the city, which radically altered the environmental characteristics of the Mar Piccolo. Today, experiments to develop farming are current, so as to diversify the production and not focus solely on mussel culture.

While *Ostrea edulis* has limited reproductive potential and is sensitive to salinity changes, *Crassostrea gigas* shows great reproductive capacity and withstands temperature, salinity, oxygenation and pH changes. These features allow easy colonization of lagoons and estuarine environments everywhere, slowly but surely replacing the native oysters. It is obvious that with the time, farmers will prefer and select *C. gigas* and today, thanks to strong imports from Japan and South Korea, two of the main producing countries with China, this species is widely present in France and Italy.

The early oyster farms in Europe

In Europe, throughout the Middle Ages, oyster consumption remained a local issue. Only at the end of the 15th century, an organized trade began, although the difficulties of transportion and storage favoured only the demand in the main royal courts, especially in France. There is a famous story of François Vatel (1631-1671), chef and master of ceremonies of the Prince of Condé, told by Marie de Rubutin-Chantal, Marquise de Sévigné (1626-1696) in her letters: the famous chef committed suicide for dishonour and shame, because a basket of fish and oysters did not arrive in time for a banquet for Louis XIV at the Château de Chantilly. So, it is not a coincidence that modern oyster farming was born in France, but only towards the end of the 19th century. The demand for oysters became more and more important and at the same time, the natural habitats of the "flat" oyster along the Atlantic coasts of France and south England were completely destroyed and fishing practically exhausted.

In the 1850s, Napoleon III had commissioned the French zoologist Victor Coste (1807-1873) to assess the possibility of farming oysters along the Atlantic coast. In 1853, Coste visited the south of France and especially the south of Italy, where, thanks to the fishermen of Fusaro Lake near Naples, rediscovered the ancient Roman technique of farming them. It consisted of dipping pottery shards, bundles of broom and ropes, mainly near small estuaries, to encourage the settlement of small oyster seeds which were then transferred to suitable habitats and protected.

These techniques had been passed down for centuries in Puglia and Campania, and rediscovered in the Bourbon period: in the mid-18th century, Charles Bourbon III, King of Naples (1716-1788), reorganized this farming technique, using seed from Taranto. In these years, Naples was the second largest city in Europe after Paris regarding number of inhabitants, and one of the most important cultural centres of the continent: according to French tradition, oysters were not absent at its court's tables. Following his father, Ferdinand Bourbon IV (1751-1825) extended the plantations around his hunting lodge near Lake Fusaro, for many years an important centre for international meetings, built by the famous architect Carlo Vanvitelli in 1782 (Fig. **86**).

Figure 86: Oyster plantations around the hunting lodge of Ferdinand Bourbon IV in the Fusaro Lake (Naples), at the end of the 18th century.

With the techniques learned in Naples, Victor Coste began a series of experiments in the Bay of St. Brieuc in Brittany and also in the Mediterranean. He obtained great results along the Atlantic coast, using wooden tiles, and then clay tiles, as collectors for seed. The immersion of these substrates at the level of low tide and in the period immediately prior to oyster reproduction allowed the harvesting of an infinite number of settled larvae. After eight months, they were detached from the artificial substrates and allowed to grow at least 2-3 years on the natural beds, before being placed on the market.

In May 1868, in the same period in which Coste started his experiments, the cargo vessel *Morlaisien* threw a gone-bad load of "Portuguese" oysters (*Crassostrea gigas*) into sea in the Gironde Estuary, but not all were dead and, finding favourable conditions, the surviving oysters were able to create natural beds across the English Channel and into the North Sea, supplanting the native oysters. From an ecological point of view, the damage was significant, but economically it was a blessing for the French farmers. At the end of the 19th century, France became the nation's leader in the production of oysters and its model was copied in southern England, as well as along the coasts of North America where in the 1920s the "Portuguese" oyster was imported. Moreover, at the end of the 19th century, the British and French oyster farmers imported large amounts of the American eastern oyster, *Crassostrea virginica*, to increase production, to aid the banks destroyed by fishing and decimated by devastating epidemic events,. These attempts failed repeatedly, but with disastrous consequences: these imports did, in fact, cause serious damage to the environment due to the involuntary import of a small but fatal mollusc, the slipper limpet *Crepidula fornicata.* It was discovered for the first time, between 1887-1890, along the coasts of Lincolnshire, and within a few decades, it spread rapidly along the European Atlantic coast, signalling the beginning of a real ecological disaster. This species, in fact, led to a radical change in the structure of the seabed, interfering with the settlement of oyster larvae and thereby preventing traditional sea farming in many areas, such as in the Gulf of Normandy (Blanchard, 1997). To add insult to injury, during the same period, together with the slipper limpet, the oyster drill *Urosalpinx cinerea*, a voracious predator of bivalves, endemic of American coasts, also arrived on the Atlantic shores of Europe around the same time (**Chapter 12**).

In addition to these problems, already very serious, very serious epizootic epidemics in 1920-21 destroyed the natural beds, forcing farmers to import increasing quantities of the Portuguese oysters, which would in turn be destroyed, in successive waves between 1966 and 1974, from other epidemics, and with which was necessary to cope with the importation of continuous stocks of *Crassostrea gigas*, mostly from Canada.

The incident of the *Amoco Cadiz*

On 16 March 1978, Amoco Cadiz, *a tanker of 234,000 tons, 330 meters long and carrying 230,000 tons of Iranian crude oil, ran aground in prohibitive weather conditions along the coast of Brittany. Waiting for help, the ship broke into two, spilling its load at sea. The environmental and economic destruction were enormous: it took more than six months to clean up 400 km of coastline, approximately 9,000 tons of oysters were lost as well as an inestimable damage to the local fauna and flora. The total damage was estimated at 250 million $ US in both tourism and fisheries. The French government asked from the insurance companies a sum of $ 2 billion.*

Modern techniques

In France, seed collection occurs during late winter using different methods depending upon the experiences of each farmer. Normally, the supply takes place by reaping the juveniles from the natural beds with special plastic instruments that favour the settlement of young larvae. In nature, *Ostrea edulis* reaches sexual maturity at temperatures not lower than 10° C releases gametes at a temperature above 14-16° C. Fertilization occurs within the pallial cavity and the larvae, after an incubation period of 8-10 days, are released free into the sea, where, after 8-15 days, they look for a suitable substrate to settle. In *Crassostrea gigas*, on the contrary, maturation, when the temperature drops below 15° C, is only delayed while the fertilization is external. The larval life is approximately 15-28 days, depending upon temperature and food availability. In hatcheries, on the contrary, the adults are kept at a temperature of around 20° C and the gamete deposition is induced by heat shock, by raising the temperature quickly to 28° C. The planktonic larvae are fed with unicellular algae and metamorphosis takes place after approximately 10-20 days: the larvae cease to float and begin looking for a support on which to definitively settle and grow. Reaching a size of 2 mm, the new settled juveniles can leave the hatchery, but they are still too small for fattening, staying for a few months in the nursery, and protected from predators. After a period ranging from 6 to 12 months, depending on growth and cultivation techniques specific to each basin, the young oysters are manually detached from the substrate in a very delicate operation that may result in the loss of up to 25% of the settled population. From that moment, the small oysters are laid out on tables elevated from the bottom, protected by nets to avoid predation, especially by gastropods, starfish and crabs. During growth, the oysters are continuously cleaned, to prevent excessive sedimentation. After two years, they are ready to be placed on the market. This type of farming, called *à plat*, is practiced primarily in Brittany, in the bays of Quiberon and Cancale, in environments subject to strong tides.

Portuguese oysters are coming from 500 kg/100 m^2 in their second year, and reach 700 kg/100 m^2 for the adults. With the "flat" oyster, they only reach 50-60 kg/100 m^2 in the second year, 100-120 kg/100 m^2 on the third and 300-450 kg/100 m^2 the fourth. However, after some epidemics, it was preferred to reduce the density of oysters per square meter, up to five times less (100 kg/100 m^2 the end of cycle), improving the final yield.

Another farming technique, called "in suspension", is practiced primarily in the Mediterranean, where the tide is virtually absent. In general, the young oysters harvested by the collectors or produced in nurseries, are placed in cylindrical close-knit containers, said "lanternet", 1 m in length and 50 cm in diameter, divided into five shelves with hanging wire mesh to about 50 cm from the bottom. The initial density does not exceed 40-50 individuals per shelf: after 16 months, the oysters have reached 45-60 mm in diameter, after 3 years an average size of 8-9 cm, with maxima of 12 cm. This technique improves the speed of growth and the quality of the product, and facilitating the maintenance. Mortality remains low because predators are more easily controlled. The “lanternet” are left immersed for approximately 10 months in the fattening stage, and just after this stage are periodically exposed to air in order to limit the development of fouling and for routine checks. In fact, at the end of spring, all the “lanternet” containers are encrusted by a set of benthic organisms, known as fouling, which can create serious problems to the yield. The fouling, in fact, competes with the bivalves for space and food, reduces the water flow through the cages, causing an accumulation of metabolites and pathogens. In addition, the weight of the containers increases, creating problems. Finally, growing on the oyster shells, fouling interferes with the vital functions of the mollusc and it certainly reduces its commercial value. In general, farmers remove it periodically with a pressure washer or leave the containers out of water for some time, even 24 hours: bivalves stand up well in the “open air”, while the majority of other organisms die. Recently, farmers have attempted to contain the development of fouling using sea urchins, efficient grazers that, when introduced into the cages of oyster farming, can limit the fouling growth.

The triploid oyster

Even in the oyster world, researchers try to select populations with special genetical features that make molluscs more robust to changing environmental conditions and to pests. In France, populations of triploid and tetraploid individuals, crossing them with wild populations, of course carried out in laboratories. The specimens thus obtained are sterile, but show a double rate of growth compared to that of natural populations. Also, these oysters do not develop gonads and so can be grown all year round: they are in fact called "four seasons oysters". On the market, however, these oysters are struggling to establish as the farmers do not take kindly to a product for which it would be obliged to buy the seed from producers, when they can pick up the newly settled larvae directly from the environment.

They also fear that these products, genetically modified, miss those features of "authenticity" that sets the original oysters apart.

Mussel farming

According to tradition, even mussel farming (*Mytilus edulis*) was born in France, but only in the 13th century. The legend tells of Patrick Walton, a shipwrecked Irishman who, in 1236, lived on the Isle of Aunis, near La Rochelle (Figuier, 1866). He noted that small mussels would settle easily on poles and on submerged nets stretched to catch fish and seabirds, thinking to breed them using ropes suspended from poles stuck in the bottom, in the tidal zone. He started collecting mussels, and the idea was a resounding success: today, just along the coast in the region of Charente-Maritime, there are more than 300 km of coasts with lines of poles to farm *Mytilus edulis*. Originating in France, mussel farming would spread rapidly throughout Europe and become especially prevalent in Spain.

In the Mediterranean, on the contrary, farmers would devote themselves to the Provencal mussel, *Mytilus galloprovincialis*, adapting the northern Europe techniques to the Mediterranean environment. Both European mussel species have had great commercial success in Europe which produces more than 600,000 tons/year, with a value of over 350 million Euros (Fig. **87**).

In recent times, and perhaps even in historical times, the European mussels were introduced to the north Pacific, China, Japan, and Australia: *Mytilus planulatus* (Fig. **88**), the subject of major farms in South Australia, with an annual output of more than 600,000 tons, is genetically very similar to the Mediterranean *M. galloprovincialis*, from which certainly originates.

Another mytilid of great commercial interest is *Perna viridis*, the green mussel (Fig. **88**), widely distributed in the Indo-Pacific region from Japan to New Guinea, from the Persian Gulf to the south Pacific. According to the FAO, more than 20,000 tons are harvested every year from natural beds, while 70,000 tons are retrieved in farming facilities. On the contrary, *Perna perna*, the brown mussel, is most prevalent in tropical and subtropical Atlantic Africa and South America where it is successfully farmed (75,000 tons/year) thanks to a spectacular growth rate: it reaches a commercial size of 6-8 cm in 6-7 months. Finally, in the last 25 years, New Zealand has also developed the farming of an endemic species, *Perna canaliculus*, the New Zealand green-lipped mussel, reaching a production of 70,000 tons in 2008.

Figure 87: Cooked blue mussels in a European supermarket.

Figure 88: *Mytilus planulatus*, the Australian mussel with an annual output of more than 700,000 tons (left). *Perna viridis*, the green mussel (right), which reaches a production of 90,000 tons/year.

In the Mediterranean, mussel farming had already developed during the Greek colonization, probably in Apulia, but there are no reliable data. It is certain that during the 18th century, the Bourbons produced mussels in the Gulf of Taranto as well as in the waters of the Phlegrean Lakes, near Naples, where, however, this type of farming could scarcely take off until the 1920s. In La Spezia (Ligurian Sea), after having tried in vain to develop oyster farming, local farmers built the first mussel farm in 1887-1888, according to the reports by Arturo Issel (1842-1922) and Davide Carazzi (1858-1923), who indicated the Gulf of La Spezia was a particularly suitable area (Consorzio Ostricoltori Spezia 1914).

In Italy, shellfish farming is the leading voice in national aquaculture, and the production is based almost exclusively on *Mytilus galloprovincialis.* With this mussel, Italians began to devote themselves seriously in the middle of the 1600s, especially in the Mar Piccolo of Taranto (Cecere and Mellea, 2009).

Farming techniques are simple and are developed according to two main systems, fixed poles or floats. The technique of fixed poles (called *bouchot* in France) is practically unchanged for centuries (Fig. **89**). Some materials changed (galvanized iron and concrete have replaced the wood, nylon replaced ropes of vegetable fibres), but the structure remains traditional. A fixed system is composed of poles driven into the bottom and placed at a distance of approximately 5 m from each other, projecting from the surface of the water to nearly 1.5 m. The poles are joined together by ropes, on which are placed the so called "reste" in Italian, nylon tubes up to 4 m long that descend towards the bottom, formed by two nets placed one inside the other, where mussels are placed. Naturally, with the implementation of the system, it is important to know the trend and the direction of the currents so as to avoid that the "neste" "do not shadow" each other, thus decreasing the intake of organic substance carried by currents. During the farming cycle (ranging from 10 to 15 months) and at regular intervals, mussels are periodically re-bagged in "reste" of increasing mesh sizes, until they reach a size suitable for sale, usually by the following summer. After three months from the start of the installation and then every 45 days, the "reste" are exposed to air for 24 hours to block the formation of fouling, to remove parasites and any dead mussels. In general, this type of farming has a low yield: in the Gulf of La Spezia production can reach about 12 kg/ m^2.

On the contrary, in floating installations (the so-called mussel longline), floating plastic buoys are anchored to the bottom and support the "reste". The system remains half-submerged, consisting of rows of variable lengths between 800 and 2,000 m parallel to the coast and placed at depths of 10-15 m. These systems can also be implemented in less sheltered areas, so exploiting much greater areas of space than those available using poles.

Thanks to the availability of space around them, the cost of work is reduced, and the yield is higher than the traditional technique, achieving 25 to 35 kg/ m^2.

Mussel production in Italy (approximately 170,000 tons/year) has stabilized in recent years after the downturn in 2003, when an exceptional heat wave hit Europe and the subsequent drought brought about a serious crisis of productivity. Mussel farming in Taranto (Cecere and Mellea, 2009), the main production area in Italy, is an important source of income for the local economy: today, the Mar Piccolo produces approximately 30-40,000 tons/year, to a value of approximately 13 million Euro, while the La Spezia production (in the north Italy) ranges between 4,000 and 6,000 tons/year.

Figure 89: Mussel longline system at Vigo (North Spain).

The Mediterranean mussel cycle

In the Mediterranean, the reproduction of mussels occurs from March to June. Fertilization is external and planktonic larvae (veligers), after 30-40 days, settle

easily on both natural and artificial substrates near the surface. The harvesting of the newly settled mussels takes place in the middle of summer, using old ropes (libani), *as collectors placed at a depth of 30-50 cm between nurseries, ropes that can accommodate from 1,000 to 4,000 individuals per linear meter.*
After the settlement, when the small mussel reaches a length of 1-2 mm and becomes a greyish colour (this phase is called "peocio", in Venice), the growth phase starts. At this point, the small individuals are inserted into the collectors ("neste") of 4-6 cm in diameter with a mesh of an appropriate size and placed at regular intervals. The 2nd cycle is usually 3-5 months later and is dependent upon the accretion occurring during the winter. The time of re-engagement takes place by April. From this moment on, the mussels should have a very rapid growth to reach commercial size (45-70 mm) by the summer.

Clam farming

Different clams are farmed around the world: the Philippine clam, *Ruditapes philippinarum*, has had great commercial success all over the world (3 million tons/year), as well as the North Atlantic *Mercenaria mercenaria*, and the Pacific *Venerupis japonica*. Another clam intensively farmed along the Pacific coast is the American *Protothaca staminea*, that was part of the diet of local ethnic groups, in particular the Chumash people, one thousand years ago. Another important clam, farmed today in Asia, is *Meretrix lusoria*, while *Mactra veneriformis* is abundantly farmed in China.
R. philippinarum, accidentally introduced in the US with *Crassostrea gigas* seed, was subsequently imported to Europe and introduced into shallow areas and brackish waters such as estuaries and coastal lagoons in Spain, France and Italy. The commercial yield, thanks to its rapid growth rate and resistance to salinity and temperature changes, as well as to low concentrations of dissolved oxygen, has been extraordinary. In Italy, it was introduced in 1983 in some valleys of the Po River (Caleri, Walloon, Scardovari and Goro), lagoons (Mistras and St Antioco) in Sardinia; in the lagoon of Marano and in the Po delta the success is particularly striking. Just seven years after planting, the "Philippine clam" supplanted, in

quantitative terms, the local species (*Tapes decussatus*), in the same high-density environments (Cesari and. Pellizzato, 1985). The success was amazing: in the early 1980s, nearly 1,000 tons/year of clams were commercialized in Italy, partly imported from North Africa, Turkey and Greece. Today, the Adriatic Sea annually produces more than 60,000 tons/year, occupying 4,000 employees to fishing that farming (source: Eurostat). Quickly, Italy became an exporting country.

The purification systems

Mussels, like most bivalves, are filter feeders: if reared in polluted waters, there is a very high likelihood that pathogens, heavy metals and other toxic substances can accumulate in their filtering systems. They must then undergo a series of good practices before consumption, especially if consumed raw. In August 1973, in Naples, large quantities of mussels, reared in highly polluted water, were placed on the local market, causing several cases of cholera. In response to this outbreak, the Italian government passed a series of laws to regulate the farming and marketing of these bivalves, requiring sanitary control during all stages of the process. Many plants, placed at a distance deemed at risk from sewage outlets (less than 500 m) were dismantled, and it was determined that the mussel cannot have a bacterial load higher than 4 *Escherichia coli* per millilitre. To enforce such provisions, depuration centres were created where mussels and bivalves in general, before being sold, are maintained for at least five days in tanks with filtered seawater purified with chlorine dioxide, ozone or UV. These plants are equipped with systems for the recovery of the disinfectants. The same disinfectants have to be used carefully because they also can cause serious environmental impacts: hypochlorite, for example, commonly used in the past because it was very cheap, reacts with the dissolved organic matter in the seawater to form chlorides, which are potentially carcinogenic. Today, it has been replaced by ozone, which has an equally strong oxidizing power, but does not produce toxic wastes. After treatment, and before being placed on the market, the bivalves are then packed in bags with traceability data and packaging date, to guarantee a safe product.

Mediterranean octopus farming

Cephalopod farming practically is non-existent, excluding some experiments for the Mediterranean octopus. The common octopus, *Octopus vulgaris*, is a commercially important species for southern Europe. For some years, it is also an important candidate for aquaculture. In theory, it should be easy, knowing some of its characteristics: the octopus adapts, in fact, easily to captivity, has a high rate of growth (at 20° C, it increases its body weight by 2-4% per day), shows a high fecundity rate, and a rather short life cycle (less than two years). In other words, a Mediterranean octopus reaches 3-4 kg of weight in 18-20 months. Unfortunately, the results have not been encouraging so far: captive farming is feasible, but research has not yet solved the problem of prey to offer as food to its paralarvae, the early planktonic development stage, and a high mortality affects this stage. Some housing attempts have also been made in Italy with the Mediterranean cuttlefish (*Sepia officinalis*). Usually, egg masses are collected in nature and placed in protected environments, to limit predation and reduce juvenile mortality.

Snail farming

Some land snails, such as *Helix pomatia* (Fig. **90**) and *H. aspersa*, have been an important source of protein for humans and their roasted shells are often abundant in late Pleistocene and Holocene archaeological sites (8,000 to 4,000 BC) throughout the Mediterranean region. These shells represent the remains of prehistoric meals and according to some archaeologists (Fernández-Armesto, 2002), land snails were the first domesticated animal (!). According to Pliny the Elder,

Fulvius Lippino began to farm them in Tuscany, Italy, in special enclosures, feeding them with a bran mixed with fragrant herbs, syrup, bay leaves and then cooked in wine, around 48-49 BC. Even Varro (116-27 BC), in his *De Re Rustica*, accurately described these early farms, while the first recipes appeared in *De Re Coquinaria*, a collection in ten books compiled around 385 AD, progressively enriched by various authors and initially attributed to Marcus Gavius Apicius (25 BC - 37? AD), the first Empire Roman gastronomist. In this collection, called *Codex Apicianus* (Battiato, 2012), the main source of information on the ancient Roman culinary habits, the author suggested to purge the snails in milk for several days. The importance of snails in the kitchen is also underlined by the fact that the Italian word "cucchiaio" (spoon) derives from the Latin *cochlear*, used by Romans to eat eggs and snails, and which was often made with mother-of-pearl. Snails soon became an important product of Roman gastronomy and in imperial time's large snail gardens called *cochleria*, were realized to farm snails using the recipes suggested by Fulvius Lippino. These "gardens" quickly spread throughout Europe and was maintained at least until the Middle Ages.

In Europe, gastronomic interest has been slowly directed toward *Helix aspersa*, known as the "petit gris" (the small gray snail) and *H. pomatia*, the "white snail", also called "gros blanc" or "Burgundy snail", two species easy to farm. At the beginning of the 19th century, they became a "must" in the French cuisine (and therefore in the world), a delicacy, and the recipe for the "bourguignon snails" would spread throughout the world by the middle of the century. Even in the Tropics, several species of snails, including *Achatina fulica* and *Pila scutata*, large land snails common in mangroves and rice fields are actively farmed. *Achatina*, in particular, a species native to East Africa and reaching a considerable size, is reared in large parts of tropics and even for several years, indoors in Europe

(Runham, 1991). This species is very prolific and aggressive and, if not kept under control, can cause serious damage to local agriculture (**Chapter 13**).

Today, the snails are considered a delicacy and are eaten traditionally at certain times of the year: generally the wild ones are consumed at the beginning of winter, when the snail goes into hibernation, sealing the opening of the shell with mucous bung which hardens rapidly, since before entering hibernation, the snails are obliged to clean themselves and their meat is not toxic. If they are harvested while still active, they should be purged in special baskets before being consumed, because may have ingested poisonous mushrooms or plants, such as hemlock (*Conium maculatum*), aconite monkshood (*Aconitum napellus*), and belladonna (*Atropa belladonna*), whose active substances are toxic to humans. Considering that the snail is basically herbivore, its organoleptic quality will depend very much on the eaten vegetation: the taste and gastronomic value can, therefore, vary greatly from area to area. The snails living in the mountains are particularly appreciated, as the Alpine specimens are larger and have a white meat which is very tender and aromatic.

Approximately half a million tons of snails are consumed worldwide each year. Almost all farms adopt a full biological cycle and are important agricultural realities in different countries both in Europe (Italy, France, and Spain) and in tropical countries. The farming of snails can be conducted outdoors on calcareous grounds, rich in water, with fences of 50 m^2, or in greenhouses, the latter system adopted mostly in France and Spain. In this case, the production costs increase, but the snails are much less subject to predation by insects, rodents and birds. The farming techniques are not difficult: snails are hermaphrodites with internal fertilization, and mating periods can last for several hours. After 10 days, both snails lay hundreds of eggs in holes in the ground of approximately 3-5 cm: after an incubation period of approximately 20 days, if the temperature remains around 18 °C, 4 mm long juveniles are born. In nature, a snail needs about 18 months to become adult, having a long period of hibernation. In controlled conditions, where the temperature never drops below 18 °C, the time is becoming shorter, up to just three months. The deposition usually takes place in early March, and the harvesting begins in October, followed by a purging period.

Figure 90: The “white snail”, also called "gros blanc" or Burgundy snail, *Helix pomatia*.

CHAPTER 12

Import and Export: Successes and Flops

Abstract: In the mid-19th century, European and North American shellfish farmers began the search for "new" bivalves of potential commercial interest, suitable for farming and more resistant to epidemics. Some of these imports have had great success (like the Japanese oyster or the Philippine clam), but often this kind of trade has caused serious problems: the introduction of a species in a "new" environment is, in fact, a high ecological risk action. It is, in fact, clear that the continuous transport of stocks of living organisms from a production area to another, practically all over the world, favours the spread of pests and the diffusion of invasive species which may cause damage to economic activities and in particular to aquaculture. Moreover, in the Mediterranean Sea, the Suez Canal has facilitated the spread of numerous molluscs which, over time, have created stable Mediterranean populations, entering sometimes in competition with the native flora and fauna.

In freshwaters, the most famous case is the diffusion in Western Europe and North America of the zebra mussel, an invasive bivalve which causes very serious economic problems, obstructing industrial and civil pumping water stations in lakes and rivers.

Keywords: *Anadara inaequivalvis, Brachidontes pharaonis, Dreissena, Crepidula fornicata,* epizootic diseases, export, import, invasive molluscs, Lessepsian migration, *Potamocorbula amurensis*, quagga mussel, *Rapana venosa*, zebra mussel.

Riccardo Cattaneo-Vietti, Egidio Trainito and Mauro Doneddu

Whenever we introduce a new species, it's an ecological roulette with nature... We'd rather not play that game.

James T. Carlton (1993), *Williams College, Massachusetts*

Molluscs have often travelled with Man: the Roman soldiers introduced for food, probably as early as the 1st century AD, a snail (*Helix pomatia*) in the UK, while the Portuguese imported an oyster (*Crassostrea gigas*) from Philippines in the 15th century.

In the mid-19th century, the trade increase of edible molluscs due to better transport systems favouring the introduction of non-native species, i.e. those not naturally present in a given environment, in both Europe and America waters. But along with a few species introduced voluntarily, many others have been able to expand, out of control, to reach new seas, lakes, and lands using human activities, especially commercial transport and navigation. Their spread is continually thwarted by Man: some of these have become pests, creating environmental and economic problems. For example, in the US National Parks, all foreign boats are inspected by NPS rangers as some freshwater invasive species, such as the New Zealand mud snail (*Potamopyrgus antipodarum*), can cause irreversible damages to the park ecosystem. Most of the alien species, however, do not usually have any reproductive success in the new environments, and fail to create stable populations, therefore being reported rarely. Often, they are not recorded again. Although not all species that Man moved more or less voluntarily on earth have caused problems; the introduction of a species in a "new" environment, as James T. Carlton suggested, is always a high ecological risk action, which absolutely must be carefully considered, usually better avoided.

Voluntary imports

In the mid-19th century, European and North American shellfish farmers began the search for "new" bivalves of potential commercial interest, suitable for farming and more resistant to epidemics. Some of these imports have had great success and today, for some species, it is often difficult for non-specialists to recognize the indigenous species from the introduced one. Sometimes, an entry is made without any apparent damages to the ecosystem: the North American clam, *Mercenaria mercenaria*, for example, a species of great commercial interest imported on the Atlantic coasts of Europe in the mid-19th century, created significant populations limited to a few areas of the coast of England, Holland, Belgium and northern France, where it prefers shallow waters and muddy bottoms.

Figure 91: The Philippine clam, *Ruditapes philippinarum* (left).The Japanese oyster, *Crassostrea gigas* (right), both introduced everywhere can develop successfully.

Its presence in the Mediterranean Sea is sporadic. In contrast, the most striking case was that of the “Japanese” oyster, also known as "Portuguese" oyster, *Crassostrea gigas* (Fig. **91**), and of the Philippine clam, *Ruditapes philippinarum*

(Fig. **91**), introduced in the last 50 years in all areas in which can develop successfully (**Chapter 11**).

Crassostrea gigas has shown great adaptive capacity, thus forming vast banks in most areas where it was introduced from North America to Australia, from New Zealand to Europe. In some cases, it has become invasive, competing with the local populations of bivalves such as *Ostrea lurida* in Puget Sound in Washington State, or *Mytilus edulis* in the Wadden Sea. Until a few years ago, the Japanese oyster was considered a different species from the Portuguese one, called *Crassostrea angulata*. Genetic analyses (Poutiers, 1993) have shown that the supposed "Portuguese oyster" was nothing more than a population of oyster living in the estuary of the Tagus River in the 15th century, as a result of an old, unexpected import from Taiwan. The Philippine clam was deliberately introduced everywhere. In Italy, it has quickly constituted large and stable populations in the northern Adriatic Sea since 1983. Its production increased from 4,500 tons in 1993 to 40-60,000 tons in the mid-1990s, and has led to a commercial success of great economic importance. Its development, however, has caused problems for the native clam, *Tapes decussatus*, whose populations were significantly reduced, and *Paphia aurea* and *Cerastoderma glaucum*, which showed a significant drop in production in recent years, even with the concomitant use of turbochargers for clam fishing (Pranovi and Giovanardi, 1998). Clam in the Adriatic Sea is, in fact, harvested with these hydraulic dredges that penetrate a few centimetres into the sandy bottom. The progress of the dredge is facilitated by water jets suspending the sediment: it is easy to imagine the environmental damage caused by these techniques.

Even *Mytilus edulis* and *M. galloprovincialis*, species farmed for centuries, have been widely introduced all over the world, heavily modifying the natural habitats. In some cases, this took place in historical times, so we now consider them native.

The same is happening with other mussels (*Perna viridis* and *P. perna*), only recently introduced in many tropical and subtropical environments. *Perna viridis,* native from Pacific Asia, is a mytilid of considerable commercial interest which has shown, in recent decades, a large diffusion capacity in the tropics: in many areas it came into strong competition with native species, causing significant changes in the structure of the coastal benthic communities. It arrived in 1999 in the south-eastern United States (Baker *et al.,* 2007), and has become one of the main components of the fouling in only a few years, the biological association developing on structures immersed in sea and forming very dense populations that even clog the pipelines at sea and weigh down the suspended structures used in aquaculture. The same problems may be also caused by the "brown mussel", *Perna perna*, another important mytilid for farming in Africa and South America, which has recently invaded the Gulf of Mexico, competing with the local wildlife.

Illegal travellers

Imports of molluscs are not always desired and controlled: often these operations are very risky, a real "ecological Russian roulette" according to Carlton and Geller (1993), because the ability to predict how an alien species can adapt itself to new environmental conditions and behave is always difficult, if not impossible. In some cases, the importation of non-native species has led to deep structural changes in the affected communities, even in very large areas, causing damage not only ecological, but also economical, sometimes quite considerable. In most cases, the introduction occurs randomly: many "unwanted" species travel in commercial stocks of shellfish or in the ballast waters of oil tankers and cargo ships which are then discharged in a new environment. Many parasites (viruses, bacteria, pathogens, toxic microalgae, protozoa) and predators (other molluscs, flatworms, echinoderms) can arrive with the imported stocks which can create very serious problems for indigenous populations: many die-offs in farmed bivalves were almost always linked to pathogenic forms which had been accidentally introduced. The spread and virulence of such outbreaks have also been related to the climate change (Occhipinti-Ambrogi and Galil, 2010), especially with sea temperature increases, which reinforces the virulence of the parasites: the protozoan *Perkinsus marinus*, one of the main agents of mortality of oysters along the east United States, was able to extend its distribution from Long Island to Maine due to milder winters, which favoured its spread northwards (Hofmann *et al.*, 2001).

Epizootic diseases

Some protozoans, Bonamia ostreae *and* B. exitiosa, *affecting the blood cells of the mollusc (hemocytes), are among the major cuases of epidemics, causing significant periodic economic damage to oyster and mussel beds (Bower et al., 1994). The first, reported in France in 1979, quickly spread to European farms, especially those of* Ostrea edulis, *where it caused losses of 40-60% of the farmed population.* Crassostrea gigas *and* C. virginica, *both intensively farmed along the American Atlantic coast and in the Gulf of Mexico, are much more susceptible to* Perkinsus marinus, *another protozoan parasite (Ford and Chintala, 2006).* Microcytos mackini *is another protozoan particularly feared on the Pacific coast of Canada, being lethal for* C. virginica *and* Ostrea lurida, *the Olympia oyster, as well as for the European "flat" oyster. Other serious pathogens are the protozoan* M. refringens *and several species of bacteria (*Vibrio aestuarianus, V. splendidus *and* V. harveyi), *especially when the water temperature exceeds a threshold value that in Mediterranean Sea is approximately 26 °C. Recently, much attention has also been focussed on the marine viruses and in particular on the ostreid herpesvirus-1 (OsHV-1), which is considered the main cause of some major mortality.*

Today, it is clear that the continuous transport of stocks of living organisms from a production area to another, practically all over the world, favours the spread of pests, even if very strict legislative rules seek to control the origin of the stocks and prevent the trade of individuals who come from infected areas. Let us also not underestimate the role that trade can have in the diffusion of some microalgae, which are often toxic. Along with oysters, macroalgae travel. They are less dangerous to humans, but able to grow in an abnormal way, resulting in structural changes in coastal communities, if the right conditions are present. In the Lagoon of Venice (Curiel *et al*., 1998) and in the Gulf of Taranto (Italy), some algal species of Japanese origin (*Sargassum muticum, Antithamnion pectinatum, Undaria pinnatifida, Codium fragile, Grateloupia turuturu*) or North Atlantic origins (*Ascophyllum nodosum*), probably introduced with imported stocks of *Crassostrea gigas,* were recently reported. It should be stressed that the forced coexistence between different species, which are genetically related, can certainly encourage the formation of hybrids over time, could create problems that are not predictable, such as a reduction in the reproductive potential of the native population. Of course, the most harmful species are those that may cause damage to economic activities and in particular to aquaculture.

Invasive molluscs

At the end of 19th century, when the American oyster (*Crassostrea virginica*) was imported in France and England, the slipper limpet, *Crepidula fornicata*, was accidentally imported also. They created many problems to the Atlantic oyster farming industry. Now present from the North Sea to the Irish waters, as well as the Iberian Peninsula, this limpet has reached very high densities and biomass in just a few years, completely changing the seabed morphology, compromising the structure of the natural beds of oysters and seriously damaging fishing in the Gulf of Normandy (Blanchard, 1997). Until now, attempts to limit their development with continuous dredging, has not yet given the expected results (Rayment, 2008). Many mussels are counted among marine species which have a high diffusion capacity, because they are very tolerant to environmental variations. *Geukensia demissa*, for example, is a mussel from brackish waters of east America, where can reach densities of more than 10,000 individuals/m^2. At the end of the 20s of the last century, probably due to oyster trading, it reached California and Mexico waters, where it has literally changed the structure of the benthic communities. Another mytilid of oriental origin, *Musculista senhousia*, has spread to Australia, New Zealand and along the Pacific coasts of the United States. In 1993, it has arrived in the northern Adriatic Sea (in the Mediterranean Sea), where it developed very dense populations: individuals are cemented to each other by a dense network of filaments of byssus, thus playing a major role in the loss of local biodiversity.

A bivalve which currently raises some concern is the mactrid *Rangia cuneata*, native of the Gulf of Mexico, which crossed the Atlantic probably together with stocks of oysters. Since 2005, it is increasing, being reported along Atlantic coasts of Belgium and France, especially in brackish waters, where it creates major problems to industrial plants because it tends to settle inside the cooling water ducts. There are many species that have travelled through the oil trade. *Potamocorbula amurensis*, native to Japan and the China Sea, was accidentally introduced with ballast waters in the Bay of San Francisco in 1986 where, over time, it has established large populations. Very aggressive, it is now considered the main culprit responsible for the collapse of some fishing activities because, in some areas it reached a cover capacity of approximately 95%, preventing the development of any other benthic species, according to the National Introduced Marine Pest Australian Information System (NIMPIS). It is, in fact, considered one of more dangerous invasive species producing a great environmental impact throughout the world.

Even some gastropods can cause serious damage. The eastern mud snail, *Ilyanassa obsoleta,* is a nassarid native of the North Atlantic, now present in California and especially in San Francisco Bay, where it is in strong competition with *Ceri-*

thidea californica, the native species now reduced to small populations in high salinity areas, where the intruder is unable to penetrate. *Cerithidea californica* is also threatened by another competitor, *Batillaria atramentaria*, a species similar to *Cerithium*, imported in the eastern United States from Japan, probably with stocks of *Crassostrea gigas* in the early 1930s, where it found a very favourable environment in California.

Some muricids, a family of carnivorous gastropods often very voracious with high reproductive and adaptive capacities, are among the most dangerous gastropods. In the absence of natural competitors, they may develop with numerically significant populations, causing serious damage to both natural and farmed bivalves. *Ceratostoma inornatum* is a muricid from Asia, imported on the west coast of the United States and in France in the 1990s, which today is considered a scourge by oyster farmers. The same happened for *Urosalpinx cinerea*, native from the North Atlantic which, thanks to the oyster trade, reached the North American Pacific and later also the Atlantic coasts of Europe.

Another fearsome predator of bivalves is the Japanese *Ocinebrina inornata,* imported on the eastern shores of the Pacific, where it attacks oysters and mussels farms (Ruesink and Buhle, 2009). The impact of *Littorina littorea*, the small herbivorous periwinkle common on the cliffs of Atlantic Europe, arrived on the coast of Nova Scotia, Canada in the mid-1800s, has been much less devastating: it adapted quickly to the new conditions and colonized the rocky littoral zone leading up to Virginia, producing changes in the algal communities structure (Steneck and Carlton, 2000).

Few molluscs are parasites. They all belong to the pyramidellids, a small gastropod family: one of these, *Boonea bisuturalis*, native of the St. Lawrence River and the north-west coast of the United States, has spread both to California and the Gulf of Mexico, causing great damage to shellfish.

Aliens in the Mediterranean Sea

The Mediterranean is considered one of the world's biodiversity hotspots. More than 1,000 alien marine species were introduced in the last century mainly thanks to the opening of the Suez Canal, fouling and ballast transportation along shipping routes, aquaculture, and aquarium trade. So the Mediterranean Sea was witness to the immigration of molluscs that have preoccupied and continues to cause concern to marine biologists. The Mediterranean basin is home to approximately 1,800 molluscs, but in the last 150 years, for various reasons, they have reported more than 200 non-native species (Zenetos, 2009). The majority of these, nearly 130 species, have reached the Mediterranean through the Suez Canal, settling mainly in the eastern basin (Gofas and Zenetos, 2003), while others have been voluntarily introduced for commercial purposes or in response to maritime traffic. The phenomenon may not be recent: it is possible, for example, that the vermetids, a family of sessile molluscs that can form large limestone platforms in the tide fringe, were already used by the Phoenicians and Egyptians as ballast stones and this habit has certainly encouraged their spread throughout the whole Mediterranean basin (Russo, 1999). More recently, two invasive molluscs have created serious problems for the fishing in the Adriatic Sea, where they have settled in the last thirty years.

The bivalve *Anadara (Scapharca) inaequivalvis* (Fig. **92**), originally from the Indo-Pacific has probably reached the Adriatic with ballast waters of oil tankers. Here, it found favourable environmental conditions for its development, especially when, with the eutrophication of the waters, the concentration of dissolved oxygen decreases greatly.

Anadara, using haemoglobin as a respiratory pigment in contrast to the majority of molluscs that use haemocyanin, is favoured in these conditions.

Figure 92: The Indo-Pacific bivalve, *Anadara inaequivalvis*, is an opportunistic species introduced into the Adriatic Sea, probably by shipping. This species is edible in Japan, but not appreciated in Europe.

The Japanese whelk *Rapana venosa* (Fig. **93**), a major predator of bivalves of Far East origins and spread around the Adriatic Sea (Savini *et al*., 2007) in the mid-1970s, reated other serious problems by attacking the mussel farms, its preferred prey. The *Rapana* larvae can survive in the ballast waters for long periods (Harding *et al*., 2004), so their ability to spread is huge: this whelk has also been reported in Great Britain, in east United States (Mann *et al*., 2006) and in Argentina. In contrast, many other species have colonized the Mediterranean in recent decades thanks to maritime traffic but do not seem to have created any problems so far.

Figure 93: The Japanese whelk, *Rapana venosa*, from the Far East spread into Great Britain, East United States, Argentina and the Adriatic Sea.

Lessepsian migrations

When the Suez Canal was open in 1869, two completely different environments, the Mediterranean Sea and the Red Sea, geographically separated by millions of years, were put in contact. A process of unique importance from a biogeographical and ecological point of view: never before had Man had so modified the natural environment. In reality, this is not so: it seems certain that the Pharaoh Ramses II (1297-1213 BC) ordered the construction of a canal, linking the Nile to the Red Sea, to open up not only the navigation between the two seas, facilitating also the spread of those few species which could withstand a "journey" in the Nile fresh waters. We have no reliable information as to what happened more than 3,000 years ago, and in any case the Nile waters had to be an almost insurmountable barrier for many species, while the process of migration through the Suez Canal in the last 150 years has been documented and very clear. Since the opening of the Canal, at first slowly, hampered by the initial strong salinity of the Bitter Lakes and then more rapidly, many species (up to date, approximately one thousand) have entered in the Mediterranean, in a process that over time has gradually assumed the characteristics of a real invasion, a migration named "lessepsian" in memory of Ferdinand de Lesseps (1805-1894), the promoter of the Suez Canal. Some species, over time, have created stable Mediterranean populations, entering sometimes in competition with the native flora and fauna.

Figure 94: *Pinctada radiata* shells washed ashore along a Tunisian beach.

Among these, many molluscs such as the shelled gastropods as *Cerithium scabridum, Conomurex persicus, Rapana rapiformis, Trochus erithreus, Malvufundus regulus, Erronea caurica, Conus fumigatus*, shell-less opisthobranchs such as *Bursatella leachi, Melibe fimbriata,* and bivalves like *Pinctada radiata* (Fig. **94**), *Fulvia fragilis, Brachidontes pharaonis, Chama pacifica,* and *Spondy-*

lus spinosus, often limiting their distribution to the Eastern basin, while others have spread all over the Mediterranean.
Among the most diffusive species, *Cerithium scabridum,* which has created large and stable populations along the coasts of Puglia and Calabria (Crocetta *et al*., 2010) and the small mytilid *Brachidontes pharaonis*, now present along the eastern coast of Sicily. Among the last to arrive, another small mytilid, *Xenostrobus securis*, and a squid, *Sepioteuthis lessoniana*, a commercially important species that spread rapidly in the Aegean Sea (Lefkaditou *et al*., 2009).

Invasive freshwater bivalves

Over the past forty years, sudden and alarming invasions of molluscs, especially bivalves, have also occurred in American and European inland waters and estuaries. Many species are often small and of no interest, while others may make up very dense populations causing not only a structural change in the communities in which they arrived, but also serious damage to submerged plants, such as pipelines and water intakes. A famous case is the invasion of the zebra mussel, *Dreissena polymorpha* (Fig. **95**), a freshwater bivalve from the Ponto-Caspian area, which can reach up to 5 cm in length. In the presence of a high concentration of calcium in the water, this mussel is able to reproduce itself with extraordinary speed, coming into strong competition with the local wildlife. According to the Global Invasive species Database by IUCN (2007), *D. polymorpha* is one of the 100 most invasive worldwide alien species, and noxious pest known. Its behaviour of forming clusters with many specimens fixed with byssus to any rigid substrate causes very serious economic problems, obstructing industrial and civil pumping water stations in lakes and rivers. The problem, known since the 19th century in Europe, became dramatic when this freshwater mussel reached North America in 1986, proliferating in the Great Lakes and the Mississippi basin, and causing tens of millions of dollars of damage. Able to settle even on the valves of other bivalves, it prevents their movement and, finally kills them: many populations of North American freshwater unionids are in trouble because of this process of intensive epibiosis (Schloesser and Nalepa, 1994).

Figure 95: *Dreissena* shells.

In Italy, *Dreissena polymorpha* appeared in the Garda Lake in the early 1970s, introduced from Germany and causing some alarm for the rapidity with which it has been able to spread in the Po basin (Lori and Cianfarelli, 2006).

Even a second European species, *Dreissena bugensis rostriformis,* originally from the Dnieper River in Eastern Europe, reached the Great Lakes in 1989, where it is known as "quagga mussel" and does not seem to be less dangerous (Mills *et al.*, 1996). Other bivalves causing similar problems are *Mytilopsis sallei* and *M. leucophaeata*, species very resistant to strong salinity changes, and able to adapt hemselves to the brackish water conditions. The first, originally from the Caribbean, is now widespread in India, Australia and the Far East (Bax *et al.*, 2002), while the second, native from the Gulf of Mexico, reached the estuaries of the Hudson River and Chesapeake Bay, the Baltic and Wadden Sea, more than 400 km muddy coastline from Holland to Denmark thanks to ballast waters.

Another small mytilid freshwater, *Limnoperna fortunei*, native of China and south east Asia, raising much concern in South America where appeared for the first time in the Río de la Plata in 1989, is spreading throughout the Paraná-Paraguay basin with a penetration rate of approximately 250 km per year, clinging to the hulls of boats sailing the basin and clogging water intakes of hydroelectric power plants (Boltovskoy *et al.*, 2006). Its rapid diffusion and ability to adapt to warm waters even polluted ones and those with low concentrations of calcium, make it a dangerous immigrant for the entire American continent. Another species with uncontrolled growth rates is *Corbicula fluminea*, a freshwater bivalve of south-east Asia, which spread accidentally due to the transport of boats or activities related to restocking fish in the rivers and lakes of the eastern United States, but is also present in France, Germany and northern Italy (Nardi and Braccia, 2004). The same happened for *Sinanodonta woodiana*, an unionid from eastern Asia, whose first record in Europe dates back to 1984, in the Danube basin. Later, it was found in central Europe and 20 years after, in Italy and France. It is a robust species that can withstand a high degree of pollution, and has been accidentally introduced with fish trade from the East, probably imported for farming or restocking. The larvae of these bivalves, *glochidium*, parasite the fins and gills of fish: with this technique, it can easy "travel" together with their host, from which it subsequently breaks off, colonizing sandy and muddy bottoms of lakes and rivers. *Sinanodonta woodiana* can definitely be considered an invasive mollusc in Italy (Niero *et al.*, 2003): in fact, in just 10 years after the first record, it is already present in eight different regions (Cianfarelli, 2009). The effects on river ecosystems are still unknown, but this alien certainly came into competition with the native unionids and, in particular with *Anodonta anatina*, whose populations are undergoing a sharp reduction in numbers. Its shell has a thick pearly layer, and today it is locally farmed for the production of freshwater pearls and mother-of-pearl: it is obvious that this economy will support the expansion of the invasive species. To control the development of bivalve pests, in general, hydrochloride was used, whose toxicity against many other organisms, causes, however, undesirable environmental consequences. Today, to avoid these problems, molluscicides that act primarily on the production of byssus, the adhesion substrate system of bivalves, are mainly used. In contrast to bivalves, freshwater gastropods are less aggressive, but *Potamopyrgus antipodarum*, a small (5-8 mm) New Zealand mud snail, has managed to invade Australia, Europe and North America (Levri *et al.*, 2007) in the last century. It can pass undigested through fish guts and is passively transported by fish and water fowl, which may play an important role in the dispersal of this snail over long distances. This species, having neither natural predators nor parasites, can quickly reach high densities in new habitats: more than 5,000 individuals/m^2 were recorded in the North American Great Lakes. Thanks to its highly adaptive capacity and reproductive rates (it is an ovoviviparous parthenogenetic species), it is able to tolerate a wide range of temperature, salinity, and trophic conditions, quickly establishing high density populations, and completely altering local ecosystems, endangering the food chain by outcompeting native snails and water insects for food, leading to sharp declines in native populations of many species.

CHAPTER 13

Dangerous Molluscs? Myths and Reality

Abstract: Some molluscs and their fossils entered in ancient myths and legends in the culture of many peoples. The most famous of them is the kraken, the legendary sea monster that sailors believed capable of attacking and sinking ships. Today, we know that giant squids (*Architeuthis*) really exist in the depths of the oceans, but their meeting is very rare.

If we exclude the myths, few truly dangerous molluscs exist: nevertheless, some species should be handled with extreme care, like the venomous cones or few species of tropical octopuses.

There is no known bivalve dangerous in itself, but they can, if ingested, become potential carriers of pathogens and toxins, causing intestinal disorders and intoxications. Moreover, some terrestrial and freshwater gastropods, especially in tropical areas, are intermediate hosts in the complex life cycles of many flatworm parasites of man, such as flukes and tapeworms. Absolutely innocent, these small gastropods are carriers of serious diseases such as fascioliasis.

Some marine bivalves, like ship-worms (*Teredo*), are able to drill, using acid secretions, wood, clays and carbonate substrates, whose action cannot be underestimated, because it has often resulted in serious damage.

Some land snails and slugs, when introduced into alien environments where there are no natural predators, can create real havoc. The giant African snail (*Achatina)* is a very aggressive pest. In general, many snails are harmful to agriculture, because they attack the more tender parts of plants useful to man, and they can also spread infectious diseases between horticultural plants.

Keywords: *Achatina*, *Aplysia*, *Architeuthis*, bilharzia, blue-ringed octopus, *Conus*, kraken, *Fasciola, Glaucus*, *Hapalochlena*, harmful algal blooms, *Lymnaea*, opisthorchiasis, paragonimiasis, paralytic shellfish poisoning, Scylla, ship-worm, *Teredo*, *Tridacna gigas*.

Riccardo Cattaneo-Vietti, Egidio Trainito and Mauro Doneddu

What was it, Sir? - said Flask. The great live squid, which they say, few whale-ships ever beheld, and returned to their ports to tell of it.

Herman Melville (1819-1891), *Moby Dick* (1851)

Myths and legends

In his cultural evolution, Man has always tried to interpret "weird" or unusual natural forms, difficult to explain, attributing them to the extraordinary and sometimes terrifying interpretations. In the Middle Ages, the fossils of unknown animals aroused much curiosity and, at the same time, ancestral fears: legends and myths were born which quickly took root in the culture of many peoples. The existence of dragons, for example, evil monsters opposed to saints and knights, was not only witnessed by dinosaur skeletons, but also by the discovery of so-called "dragon's teeth". In truth, these latter are rudists, large Mesozoic cone-shaped shells, abundant in the Cretaceous but now extinct. Even megalodontids, bivalves typical of the Upper Triassic, whose cross-section is shaped like a hoof, were interpreted as "a mark of the devil": the testimony of the existence of the devils themselves. In many Alpine areas, topographic names as "the Devil's Seat or the Devil Mountain", are probably linked to the presence of these shells in the earth, belonging to molluscs which lived more than 200 million years ago. In Italy, formations of this type are present throughout the Dolomites, and in particular in the area of Croda da Lago.

One ancient myth concerns *Aplysia*, large herbivorous and harmless sea slugs common inshore in tropical shallow waters and in the Mediterranean. Known by Romans as *lepus maris* (sea hares), they emit, when disturbed, an acrid and fetid purple ink. The name *Aplysia* means “indelible” and this fluid was once believed to be a poison, killing even the person who touched the animal with a stick. According to Pliny the Elder (23 -79 AD) and Claudius Aelian, a Roman naturalist of the 2nd century AD, *Aplysia* secretions were extremely toxic, capable, if ingested, of inducing vomiting, while its contact would make evident a concealed pregnancy (Fleming, 1829) or procure hair loss: the great Swedish naturalist Karl Linneus (1707-1778) 'resumed' this belief, calling one of the Mediterranean species *Aplysia depilans*, or "hairless *Aplysia*".

Indeed, the aplysiids can be toxic to marine birds, and possibly also to mammals, providing serious paralysis and lung injury (Carefoot, 1987), but they are unlikely to be dangerous to humans. However, according a legend due to John of Antioch (Roberto, 2005), the aplysiids had a very famous victim, the Roman Emperor Titus (39-81 AD), the destroyer of Jerusalem. Titus, before returning to Rome after the victory in Judea, asked to Apollonius of Tyana, diviner in Cilicia, as to who would die: his cryptic answer, as usual, had been: *you will die like Ulysses*. Knowing that Ulysses was dead at sea, Titus always kept away from sea and its inhabitants, until his brother Domitian, eager to succeed him, offered him a lethal meal of "sea hares".

In the 19th century, a popular belief was that the Indo-Pacific clam *Tridacna gigas*, a bivalve which can reach a length of more than one meter and a half, could close its valves very quickly, imprisoning the arm or foot of an unfortunate pearl diver. In the early 20th century, Alpheus Hyatt Verrill (1871-1954), a serious scientific writer and the son of the first Professor of Zoology at Yale University, was thought to believe this legend. Only after greater knowledge of the tropical seas and the spread of diving activities, it was possible to realize the extreme slowness with which the clam closes its valves, thereby avoiding any danger.

The kraken

Perhaps the most famous myth linked to molluscs is the kraken (Ellis, 1999; Williams, 2010), the legendary sea monster as “large as an island”, that sailors believed capable of attacking and sinking ships, as Melville reminded us. Its existence had always been very controversial, although the whalers of the 18th century often found huge tentacles in the stomachs of sperm whales. Science, however, needed more tangible evidences: for a long time, the Academy did not believe these findings, accusing fishermen and sailors of inventing extraordinary tales. The myth became reality only at the beginning of the great geographical explorations, at the end of the 18th century, when some encounters with these giant squids, although rare, began to be reported from the Atlantic Ocean. The descriptions became more and more frequent, and more accurate, during the 19th century and the scientific community began to take an interest in these giants of sea. In October 1873, Theophilus Piccot and Daniel Squires, fishermen from Newfoundland, caught a virtually intact specimen, and all doubt was thus dissolved. In 1883, at the International Fisheries Exhibition in London, a full-scale model of a giant squid was exposed and aroused much curiosity in the visitors who could not imagine that there were really "monsters" of that size. Japetus Johannes Steenstrup (1813-1897), a Danish zoologist, described it in 1857, calling it *Architeuthis dux,* suggesting that many other sea creatures described in the medieval and Renaissance literature including the "sea monk" by Guillaume Rondelet (1507 -1566) in *Libri de piscibus marinis in quibus verae piscium effigies expressae sunt* (1554) and by Conrad Gessner (1516-1565), were in fact giant squids. The same was also true for sightings of other monsters, such as the "sea serpent" and the "lusca," a monstrous creature, perhaps a giant octopus, reported from the Caribbean Sea which could be just fanciful interpretations of encounters with giant squid with very long tentacles. It is more difficult to connect the north European myth of the kraken with the Greek myth of the Scylla nymph, whose body also had long tentacles: the kraken is not, in fact, present in the Mediterranean although the Phoenicians, great navigators, could meet in the Atlantic.

The Scylla legend

Scylla, the daughter of the Goddess Crateide, was a beautiful nymph living near Reggio, on the Strait of Messina. Glaucus, Neptune's son, half man and half fish, fell madly in love with her, but his love was not reciprocated. He asked the sorceress Circe for help, who, in turn, fell in love with the young man. Rejected by Glaucus, faithful to Scylla, Circe took revenge by turning the poor nymph into a monster: instead of legs appeared six dogs with long necks in the shape of snakes (the tentacles of a giant octopus?) and a hideous mouth (the horny beak-shaped jaws of cephalopods?). Hidden in a sea cave, the furious Scilla began to attack any ship that had the misfortune pass through the Straits, including those of Odysseus and Aeneas.

The kraken, a squid of considerable size and weight of more than 2 tons, lives deep in the ocean and its rarity means that we do not know with any certainty how many species it comprises. For some there is one unique species, *Architeuthis dux*, but for others (Roper *et al.,* 1984) the described species, both in the Pacific and the Atlantic Ocean numbering approximately twenty, can be reduced to three: *Architeuthis dux* in the North Atlantic, *A. santipauli* from the Southern Ocean, and *A. japonica* from the North Pacific. The giant squid are less rare than we think because they are also the favourite food of the sperm whales: certainly many sperm whales show scars on their skin from the hooked suckers and giant tentacles were often found in their stomachs. However, the finding of complete individuals is still very rare: at the end of the 19th century, some specimens were washed ashore on the coast of the Newfoundland, and others, more recently in New Zealand, were in poor condition. Few museums in the world are able to show a giant squid: on the other hand, bodies of this size are certainly not easy to preserve. In 2004,

Kubodera and Mori (2005), closely following sperm whales, were able to take the first underwater images of a giant squid at 900 m deep in a canyon off Ogasawara Island in south Japan.

Today, the scientific literature describes other large squids belonging to families different from that of *Architeuthis*, such as *Mesonychoteuthis hamiltoni* that typically measures 6-8 m, and can weigh 450 kg, typical of the Antarctic waters. It must be emphasized that prudent behaviour is never enough in the presence of large squids, which are often very aggressive, a: the Humboldt squid (*Dosidicus gigas*), for example, reaches 2 m in length and 50 kg in weight and is common in the waters of California. The Mexican fishermen call it "el diablo rijo" for his hard sharp horny beak and hooked suckers on its arms are potentially dangerous and can cause very serious injuries to divers who meet it.

Some giant octopus really evoke fear and have become part of the classical myths: Pliny the Elder told about an octopus which ventured onto land in search of food at Carteia, near Cadiz in Spain, while Claudius Aelian remembered a giant octopus, usually stealing salted fish in Pozzuoli warehouses, near Naples!

At other times and in other places, the great octopus of the North Pacific and the Bering Sea, *Enteroctopus dofleini*, attaining 70 kg in weight and length of 7 m (Cosgrove, 1987) may have created the myth of the giant octopus.

Bites and stings

If we exclude the myths and incredible stories, few truly dangerous molluscs exist: nevertheless, some species should be handled with extreme care. The cones (Fig. **96**), carnivorous gastropods with a beautiful shell of great interest to collectors, can inject very dangerous neurotoxins: deaths by cardiac arrest or gangrene are known to occur after these cone stings. The German naturalist Georg Eberhard Rumphius (1627-1702), living in Indonesia, described the first case of fatal poisoning in the Island of Banda. These gastropods have, in fact, a deeply modified mouthpart and the radula, the typical gastropod scraping organ, has teeth that can inject a mixture of toxins that, at a minimum, can cause temporary paralysis of the limbs, but sometimes, in the more sensitive subjects, can be deadly: the action of toxins, in fact, causes respiratory failure and when the diaphragm and lungs are paralyzed, a cardiac arrest occurs. Many accidents are also related to the fact that cones are among the gastropods the most sought after by collectors for the beauty of their shell, and often inexperienced collectors have been poisoned by these species.

Figure 96: A living *Conus geographicus*.

Fortunately, only a few groups of species are really dangerous, of which all are present in the Indo-Pacific, and easily recognizable: *Conus geographicus* (Fig. **96**), *C. striatus, C. tulipa, C. textile, C. aulicus* and *C. marmoreus*.

Conus geographicus, being the most common, is also the species that has caused the most accidents, but *C. tulipa* and *C. striatus* are generally considered the most venomous. The small piscivorous cones, such as *Conus catus, C. magus, C. monachus* and *C. obscurus*, are of lesser hazard, but must still be handled with care. Even in the tropical Atlantic, several species of cones, *Conus centurio, C. clenchi, C. clerii, C. ermineus, C. jaspidus* and *C. regius*, are potentially dangerous, but there are no reports of fatal poisonings (Haddad *et al.*, 2006).

Figure 97: A road sign warning the presence in the waters of the "blue ring" octopus (Eastern Australia, 1986).

Apart from cones, few molluscs can be poisonous to humans: among these is the small and very rare pelagic nudibranch gastropod, *Glaucus atlanticus* and *Glaucilla marginata*. They both live in warm temperate waters (Thompson and Bennett, 1969): feeding on one of the most venomous species of jellyfish, the Portuguese Man-o'War, *Physalia physalis*, they are able to assimilate and concentrate its poison. Since these small planktonic molluscs very beautiful and eye-catching, and are often washed up in droves on the shores, it is very tempting to pick them up, but the contact can be very unpleasant as even moribund they can inject a significant amount of poison.

Among the cephalopods, the small blue-ringed octopuses (*Hapalochlena lunulata, H. maculosa* and *H. fasciata*) living in the Pacific are famous for their danger, because of their bite injecting an often lethal tetrodotoxin (Sheumack *et al.,* 1978). In this case, death occurs by paralysis of the respiratory muscles (Fig. **97**).

We must not forget that all cephalopods have a horny beak so that even the common octopus and cuttlefish can have a very painful bite and so require some care in handling. Some shellfish, finally, act as carriers of dangerous parasites, while the filter-feeding bivalves can cause allergies and poisoning, also very serious if consumed during "red tides", when the toxic microalgae are present in very high concentrations in the waters where the bivalves live.

Bivalves, dangerous filter-feeders?

There is no known bivalve dangerous in itself, but they can, if ingested, become potential carriers of pathogens and toxins, causing intestinal disorders and intoxications. Almost all bivalves are extraordinary filter feeders (a mussel filters 3-4 litres of water/hour, depending upon temperature) (Jorgensen, 1990) and are, therefore, able to gather large amounts of particles (viruses, bacteria, microalgae, etc.) in their hepatopancreas. If living in polluted waters, this bioaccumulation of pathogens, heavy metals (especially copper) and chloride derivatives (DDT, PCB, etc.) may be very high. Unfortunately, the consumption of raw bivalves harvested in contaminated waters is common in many regions and this certainly facilitates the onset of major diseases such as typhoid and paratyphoid, salmonella, hepatitis, viral gastroenteritis, and even cholera. In 1973, the people of Naples and Bari (Italy) suffered a cholera outbreak due to the consumption of raw seafood farmed in the polluted waters of the harbour, a criminal offence which can have serious health consequences.

Another source of contamination are the "red tides" (Harmful Algal Blooms, HABs), the abnormal development of toxic microalgae known as the paralytic shellfish poisoning (PSP) algae. These HABs are events virtually impossible to predict, a phenomenon typical of warm-temperate waters, and one of the most serious problems facing aquaculture.

Most cyanobacteria and microalgae are absolutely harmless, but a limited group of species may produce powerful toxic substances that accumulate up the food chain, especially when they reach very high concentrations in the water column (in the order of millions of cells per litre). It is obvious that, in these cases, bivalves are transformed into deadly accumulators and carriers of these toxins that, when ingested, can be very dangerous to human health. Not only bivalves can accumulate toxins derived from the diet: some pteropods, such as *Limacina retroversa*, were considered the main cause of serious mortality of herring in the Bay of Fundy, in the Gulf of Maine (White, 1977), as carriers of toxins of dinoflagellates to which they are apparently immune.

Molluscs, involuntary diffusers of parasites

Some terrestrial and freshwater gastropods, especially in tropical areas, are intermediate hosts in the complex life cycles of many flatworm parasites of man, such as flukes and tapeworms. Absolutely innocent, these small gastropods are carriers of serious diseases such as fascioliasis. Freshwater snails of the genus *Lymnaea,* in fact, play an important role in the life cycle of *Fasciola hepatica*, while *Melania* and *Melanoides* snails are required in the cycles of *Paragonimus westermani* and *Opisthorchis sinensis*, and *Planorbis* and *Bulinus* snails host the larvae of *Schistosoma*. All these parasites can create worldwide problems. *Fasciola* is a very serious parasite: approximately 2 million people are infected in the world, mainly in Latin America, Africa and China. In rural Europe, cases of fascioliasis are still recorded, due to *Fasciola hepatica*, occurring if raw vegetables grown in soil irrigated and frequented by herbivores are consumed.

In France, 2-6% of cattle are infested, and 70-80 human infections are reported per year. In Sardinia, the parasite affects mainly sheep, while in the Po Valley this parasite was common among cattle until the 1950s. The life cycle of the various species of *Fasciola* is complex: the eggs present in the faeces of herbivores, subsequently released into the water and the larvae, called *miracidia*, continue to infest an intermediate host, the *Lymnaea* snails, generating a second larval form, *redia,* in asexual reproduction. Under optimal conditions of temperature, *redia* produce infesting forms, *cercariae* which, when released into the water, encyst onto the leaves of aquatic plants that the herbivorous, the definitive host, ingest on eating them. Sometimes, man can replace the herbivores, becoming the definitive host. In Europe, *Fasciola hepatica* recognizes a single intermediate host, *Lymnaea truncatula*, but in other regions where it was introduced with domesticated animals, it has adapted to different species such as *Lymnaea humilis* in North America, *L. viator* in South America and *L. tormentosa* in Australia.

In the case of the bilharzia fluke, *Schistosoma* uses a freshwater mollusc as the intermediate host to complete its life. The parasite causes anaemia, diarrhoea and severe internal bleeding in the final host: more than 200 million people in 76 different countries are affected by this tropical disease that causes the death of nearly 200,000 people every year only in the sub-Saharan Africa. Also in this case, the cycle of the parasite begins with the issuance of eggs in faeces of a definitive host, typically a water bird. The larva, called *miracidium*, is ingested by pond snails (the intermediate hosts) belonging to different genera such as *Lymnaea*, *Physa* and *Planorbis*. The larva develops asexually, inside the mollusc hepatopancreas, producing the next larval forms (*sporocysts, redia*) that, ultimately, give rise to *cercariae* that leave the snail and actively go looking for their definitive host, encysting in other intermediate hosts (fish, amphibians) or on vegetation until the final host is reached. When man, taking a bath or feeding on infected plants, comes into contact with these larvae, they penetrate his skin, reaching the blood vessels and then the intestine, where they reproduce and lay eggs, and the cycle begins again. Man has encouraged the spread of these pests with his sedentary behaviour: for millennia, in certainly not ideal hygienic conditions, he lived in close contact with animal droppings, thus promoting auto-infestation processes in the presence of molluscs which were absolutely innocent.

Paragonimiasis is another parasitic disease caused by trematode flatworms belonging to the genus *Paragonimus,* localized in the lungs of several carnivorous mammals. It is widespread in Asia, Africa, and South America. The definitive host is infested by eating raw, slightly cooked or preserved in brine freshwater crustaceans, infested with these larval forms, called *metacercariae*. In this case, the intermediate hosts of these parasites are freshwater gastropods of the families Pleuroceridae and Thiaridae.

Opisthorchiasis is another form of parasitic infestation by flukes in which the role of freshwater gastropods is crucial: the different species of *Opisthorchis* affect the bile and pancreatic ducts, especially in dogs and cats, while philofthalmiasis (the *Philopthalmus* fluke), passing through a gastropod, can lurk in the eyes of waterfowl and occasionally in man. Heterophyasis is another infestation due to the intestinal fluke *Heterophyes heterophyes,* which may even settle in the brain and heart in the definitive host (Gangemi, 2012). Finally, the nematode *Angiostrongylus cantonensis* is the agent of a serious form of meningitis, and takes advantage of the land snails as its intermediate hosts, while definitive host is the mouse: it follows that it is very dangerous to eat raw snails in areas where the parasite is present, in tropical areas and in particular in south-east Asia and the Pacific.

Industrious borers of rocks and timber

Some marine bivalves are able to drill, using acid secretions from glands present in their mantle and the mechanical action of their shells, wood, clays and carbonate substrates, including the skeletons of stony corals. Species belonging to the genera *Lithophaga* (Fig. **100**), *Pholas, Gastrochaena* and *Teredo* are, especially in the warm temperate zone, the greatest exponents of these particular bivalves, whose action cannot be underestimated, because it has often resulted in serious damage. In particular, ship-worms, belonging to a small family of marine bivalves, are very harmful. They are able to dig deeply, causing considerable damage to the ship hulls, and wooden constructions such as piers and dams. The best known species is *Teredo navalis* (*teredon* in Greek means worm), but several species were described worldwide, and are not always easily recognizable (Fig. **98**). Typically only a small part of the body of the mollusc is contained by the shell, a long calcareous tube easy to see in old beached woods. Growing inside the wood with an infinite number of individuals, the ship-worm weakens the structure, which finally collapses. The Christopher Columbus caravels, and in particular the *Santa Maria*, suffered greatly from the action of these bivalves, but the whole history of navigation is full of shipwrecks caused by them. The *Golden Hind*, the famous galleon on which Sir Francis Drake (1540-1596) circumnavigated the planet between 1577 and 1580, sank at Woolwich on the Thames, England, following the action of ship-worms. In addition, according to some historians, the defeat of the Spanish Armada in 1588 was not only due to the action of English fleet, but also to these insidious bivalves that had affected the Spanish hulls which, during a storm, collapsed. Even some catastrophic floods occurring in the Netherlands during the 18th century were caused by ship-worms that damaged the wooden dams.

Figure 98: *Teredo* tube-shells inside a piece of wood.

In Venice, ship-worms are particularly feared. Probably introduced to the Mediterranean and in particular in the Venetian Lagoon in the 17th century, ship-worms constitute a serious problem especially for the protection of the pilings on which a considerable part of Venice is built. Periodically, the stakes are isolated from the environment and "painted" with a layer of tar to prevent larval attack and "choke" any individual incumbents. In the mid-18th century, it was realized that the copper was toxic to many marine species and in particular for ship-worms, so the Royal Navy, during the American War of Independence, began to cover the ships' hulls with copper sheetings. Today, the protection is entrusted to paints that prevent the drilling of the ship-worms in submerged woods.

Crop pests

Some land snails and slugs (these last without a shell), when introduced into alien environments where there are no natural predators, can create real havoc. The giant African snail, *Achatina fulica*, whose massive brownish shell can reach 30 cm, is a very aggressive pest. Native from eastern Africa and of considerable commercial interest, it was a widespread source of for food in different areas of tropical Asia and in many islands of the Indian and the Pacific oceans. In the 1990s, it was also introduced in the West Indies, where its high rate of fertility and ability to adapt to different environmental conditions made it highly invasive, while its voracity makes it very dangerous for local agriculture. It is now considered a pest in many tropical areas and its spread is fought with energy, although with great difficulty. An attempt to control the spread of *Achatina* by introducing its predator, the carnivorous snail *Euglandina rosea*, caused, in turn, more environmental disasters in many Pacific islands. This predator, in fact, often completely ignoring the expected prey, turned instead to the native snails, destroying entire populations (Civeyrel and Simberloff, 1996). In Hawaii, twenty-two endemic species of gastropods have already become extinct and the remaining nineteen are considered endangered (Wilson, 1992). Less aggressive, but with equally high diffusion capacity, is *Pomacea canaliculata*, a freshwater snail widely distributed in the rivers of South America that, in recent years has managed not only to invade the inland waters of north America, but also south-east Asia, from Thailand to the Philippines, especially in areas where rice and taro, a tuber similar to potato, are farmed. In Europe, *Arion vulgaris*, the Spanish slug, has spread across the continent in the last sixty years, causing serious damage to vegetables. Its reproductive capacity is so high as to make it ineradicable.

In general, many snails (*Arion, Limax, Helix* and *Cyclostoma*) are harmful to agriculture because they attack the more tender parts of plants useful to man, and they can also spread infectious diseases between horticultural plants (Barker, 2002). Taking into account that the snails are very prolific, they must be controlled in agricultural environments: to catch them, traps buried in the ground and partially filled with beer, which the snails love, are used; today the most commonly used baits are based on metaldehyde. A new type of biological control that is spreading in agriculture involves the diffusion in the soil of a slug parasite, the nematode *Phasmarhabditis hermaphrodite.* While some insects, such as beetles and larvae of fireflies, toads and many birds such as blackbirds, thrushes, and some ducks are very useful in the fight against harmful gastropods. It is useful to point out that in a natural environment these pests find "controllers", while in agricultural areas where, for one reason or the other, there was a decrease in biodiversity level their control becomes much more complex. In these cases, it is therefore necessary to intervene with chemicals (pesticides, molluscicides) that, in turn, cause damage to the same environment.

CHAPTER 14

Threatened Species

Abstract: Due to the heavy impact of the human activities, a generalized habitat depletion, with loss in terms of biodiversity, biomass, and structural complexity, has been recorded everywhere in rivers, coastal waters, estuaries and lagoons. Molluscs have suffered and are suffering of this pressure and in the more polluted environments only the more "robust" species are able to adapt to these conditions, but the loss of biodiversity both at sea and land has been very serious.

The exploitation of limpets, mussels, oysters or other marine bivalves, living in the coastal zone, has been significant in the last two centuries, and trawling gears, like dredges and bottom nets, caused serious structural changes in all exploited benthic communities, putting in jeopardy the existence of various habitats and many species.

Today, many molluscs of commercial interest at risk of extinction are reported in CITES appendices, the UN organisation which controls and regulates trade of wild fauna and flora, trying to prevent the exploitation of endangered species. However, controls are often lacking and very often many molluscs are still harvested indiscriminately. Other species, not yet taken into account, must be protected as soon as possible: in the fight against the loss of biodiversity, it is necessary to take actions that they will not only protect a particular species, but the whole environment where the species lives. It is also urgent the protection of the fossil deposits which are full of shells, often of great collection interest.

The establishment of parks and protected areas, and an international legislature that manages the marketing, seem today more necessary than ever for the protection of the living and fossil molluscs.

Keywords: *Bolma,* CITES, climatic changes, *Concholepas*, date-mussel, fossil shells, *Gibbula nivosa, Lithophaga lithophaga*, molluscs, *Morum,* mullu, *Papustyla pulcherrima,* pollution, *Strombus gigas*, threatened species, trawling, *Trochus, Turbo.*

Riccardo Cattaneo-Vietti, Egidio Trainito and Mauro Doneddu

Will tomorrow's child ask why we did nothing on our watch to protect the sharks, tuna, coral reefs, and the other threatened life of our living oceans?

Sylvia Earle (1935-), *National Geographic Society. TED Prize 2009 winner.*

Man has always caused serious structural and functional changes in both the terrestrial and marine environments: in many cases, unfortunately, in an irreversible way. There is no environment of the planet in which human activities have not altered the habitat structure and its populations. During the last decades, the Mediterranean rocky benthic communities up to 30-40 m depth, have shown significant changes in their structure and taxa composition, deeply modified by natural factors (including global warming) as well as human activities. Generalized habitat depletion, with loss in terms of biodiversity, biomass, and structural complexity, has been recorded in many areas (Bianchi and Morri, 2000; Coll *et al.*, 2010; Claudet and Fraschetti, 2010; Sala *et al.*, 2012; Bo *et al.*, 2014). Depending upon local conditions, the organic pollution, the increasing sedimentation and consequent water turbidity, the periodic beach nourishment, the physical disturbance by fishing gears, the impacts of recreational activities, the global warming determining mass mortality events (Garrabou *et al.*, 2009), the changes in the diffusive capacity of a certain species due to climatic changes, and the extensive presence of introduced species have all driven structural changes of the marine communities. The lost or sharp reduction of some sessile key-species among macroalgae, sponges, gorgonians, and bryozoans, as well as the emergence of others, once considered rare, often determined a significant reduction in habitat complexity (Garrabou *et al.*, 1998; 2002; Airoldi and Beck, 2007).

Rivers, coastal waters, estuaries and lagoons, these last among the most productive environments in the world, have become the receptacle of all the toxic substances discharged by man over time. Chemicals, both inorganic (heavy metals, especially copper) and organic (hydrocarbons, chloride-derivatives, DDT, PCBs, etc.), released into the environment and easily accumulative in the living tissues, have produced considerable stress in the populations, limiting their reproductive and immunological capacities. Naturally, molluscs have suffered and are suffering many of these pressures that have often led to the depletion or disappearance of some populations, if not of the species. In fact, the more sensitive populations were decimated, and only the more "robust" species were able to adapt to these conditions, but the loss of biodiversity both at sea and land has been very serious. It is almost impossible to outline in its entirety all the shellfish that has been, in some way, threatened by human activity, taking into account that since prehistoric times, Man, gathering shellfish as food, began to affect many populations, especially those of the unionids, collected in quantity in rivers and lakes (Mannino and Thomas, 2002). With the industrial revolution, pollution poisoning the waters or the construction of dams along the courses of the rivers, completely changing the habitats, determined a state of serious decline for many species: to give just one example, in the United States rivers approximately 300 species of unionids, a unique and extraordinary richness, are today in great danger. A quarter of them are seriously threatened and some species, such as the "fat three-ridge" *Amblema neislerii*, are considered close to extinction.

Many limpets, mussels, oysters or other marine bivalves, living in the tidal zone, have always been a natural source of food for many human populations on all continents as testified by the shell heaps found worldwide (**Chapter 10**). The exploitation of molluscs has always been significant, but in the past harvesting was mostly done by hand, in areas of low tide and consumption directly linked to the needs of local people. In addition, the coastline was so populated and the harvesting so limited that the mollusc populations were readily repopulated from the surrounding areas. At first, therefore, the freshwater populations suffered much more than the marine ones; however, over the past 200 years, with the use of more efficient fishing gears and the increase in water and land pollution, the risk of extinction for many species, including marine, became very real. Steam fishing boats appeared in the early 20th century, with the subsequent possibility of operating in deep waters, trawling with dredges and bottom nets, extremely destructive gears, caused serious structural changes in the benthic communities everywhere, putting in jeopardy the existence of various habitats and many species.

The natural in-shore oyster beds in almost all marine and estuarine environments of the temperate-cold areas were destroyed at the end of the 19th century by a reckless fishing. The damage was not only economic: these banks had a role to play in the protection of the coastline from the waves and tides, and their disappearance has been considered in many areas the main cause of littoral erosion on both sides of the Atlantic coasts. In the 19th century, the city of Biddeford, Maine, forbade even the harvesting of mussels around the pylons of the bridge over the Saco River, as mussels not only formed an effective protective barrier against the wave erosion, but occupied the spaces between the stones and reinforced the pylon itself (Verrill, 1939). Moreover, we must also consider that once a shelled mollusc is dead, its shell contributes to the natural nourishment of the beach: the coastal erosion, a major problem in many littoral areas, would probably not have the same gravity today if the natural beds of bivalves were still present.

A good example of the devastating effects of the industrial revolution on molluscs is the fate of the Olympia oyster banks, once a staple food for the people of California and Oregon. During the "gold rush" (1848-50), these banks were completely destroyed not only by overfishing, but also by the huge amount of mud due to the very coarse processes of mining. Today, restoration programs conducted in the San Francisco Bay, Puget Sound, and Oregon are trying to replenish those banks, lost more than 150 years ago. The same devastation occurred even in the tropics: the continued destruction of the mangroves that line the continental coasts between the isotherms of 24 °C, to achieve coastal areas suited to the farming of shrimps or even rice paddies, caused possibly irreversible damage. The various methods of fishing, including dynamite and poison, conducted both for food or collection, led to a rapid decline of many populations of bivalves and gastropods. Today, many molluscs of commercial interest at risk are reported in CITES appendices (*Convention on International Trade in Endangered Species of Wild Fauna and Flora*), by the UN Agency for Environmental Protection which controls and regulates trade of wild fauna and flora (or products thereof) in danger of extinction in the world, trying to prevent the exploitation of endangered species (Wood & Wells, 1995).

The CITES Organization

The Convention on International Trade in Endangered Species of Wild Fauna and Flora (CITES) aims to protect plants and animals considered endangered or at least threatened by human activity, regulating and controlling their international trade. The Convention entered into force in 1975, currently supported by 175 countries. In 2009, CITES regulated international trade in some 30,000 species, of which about were 25,000 plants. The list of these species is listed in three appendices available on Internet. The first shows the species that are seriously endangered: they must not be harvested, neither marketed. Appendix II brings together the species whose existence is not directly threatened by the man, but are considered at risk: there are no restrictions in their trade, but marketing is seriously monitored. Appendix III, includes those species that, at the request of a member country, appearing very or illegally exploited.

Most shellfish for which trade is prohibited or strictly controlled live in freshwaters, such as the unionids, a family at high risk of extinction due to pollution. Among the marine species, many large tropical gastropods are under CITES control, such as *Strombus gigas*, whose harvesting is regulated, several species of *Cymbium*, now sold as dried meat even on the Internet, *Trochus* spp., harvested also to extract mother-of-pearl, *Bolma girgyllus,* and different species of *Turbo* and *Morum*, often of considerable interest for collectors.

In South America, the "loco", *Concholepas concholepas*, a gastropod with a high gastronomic value in Chile, is in serious distress: in the past 40 years, its exploitation has been heavy. In 1975, approximately 5,000 tons were collected, surpassed in 1980 to 25,000 tons.

Following the collapse of natural populations since 1989, its fishing is now controlled (as a minimum size of 10 cm and seasonality) and since 1992 permission is granted only to a predetermined number of professional divers. The economic interest of the species is very high and an attempt to farm it was done, with little success.

In South America, the intensive harvesting of some species of thorny scallops, the mythic *mullu*, food of the Gods according to a Quechua myth, has virtually reduced to zero the local population in the Gulf of Guayaquil in Ecuador. Today government recovery programs have been activated to prevent the disappearance of this species, a symbol of the pre-Columbian culture (Fig. **99**).

In the Far East, even the natural populations of the Oriental common clam, *Meretrix lusoria*, are at risk in the waters of Japan.

Fishing and pollution are, therefore, primarily responsible for the progressive depletion of shellfish populations. The spread pollution of coastal waters has certainly provoked the decrease if not the disappearance of many species. During the mid-19th century, an Italian malacologist, Salvatore Trinchese (1836-1897) collected all the known Mediterranean species of Sacoglossa (a small order of sea slugs) in a few years within the Port of Genoa; today, of course, it would not be possible even along the Ligurian coastline, given the changed environmental conditions. This process of loss of local biodiversity must be fought against as much as possible, although its importance escapes the attention of most.

Figure 99: Ecuador government advertisement to help save the over-exploited *Spondylus* populations, the *mullu* for the Quechua population.

Synergistic processes between chemical pollution and bacterial or viral epidemics, often for unknown reasons, affected many populations: in the early 1980s years, most of the Mediterranean populations of *Spondylus gaederopus*, the thorny oyster, were dramatically reduced and are still struggling to recover after a few decades. The reasons for the die-off has not yet been clarified, although recently have been isolated several strains of pathogenic vibrios in marine invertebrates.

A secondary cause to the depletion of some species is also the harvesting to supply the tourist market, for souvenirs, food, jewellery or simply, to supply the craft that produces gadgets and, in tropical areas the "puka shell jewellery". Other species are protected because they are actively sought for their interest to collectors and as a result have become rare. For example, a terrestrial snail, *Papustyla pulcherrima,* with a spectacular emerald shell, endemic to Manus Island (Papua New Guinea), has been actively collected in the last decades, so that it is now on the list of endangered species and its trade became illegal.

In the Mediterranean Sea, *Mitra zonata*, *Zonaria pyrum*, *Luria lurida*, *Erosaria spurca,* and *Schilderia achatidea* are all species of great interest to collectors, actively sought and therefore potentially at risk. Of particular interest is a small endemic gastropod, *Gibbula nivosa*, known from only a very narrow distributional range: alive it has been found only in the Bay of St. Thomas in the Island of Malta, where it is obviously protected.

The date-mussel: a Mediterranean story

*In the Mediterranean Sea, the harvesting of the date-mussel (*Lithophaga lithophaga*), a bivalve of great gastronomic value, is forbidden in many countries because its removal requires the dismantling of the carbonate rocks in which the bivalve lives, producing an irreparable damage and the total alteration of these benthic communities (Russo and Cicogna, 1992).*

In addition, it also causes damage to coastal stability and a permanent change in the underwater landscape (Guidetti and Boero, 2004). Today, the date-mussels are protected by the EU and their harvesting and consumption banned in Italy since 1998 (Fig. ***100****).*

Some Mediterranean large bivalves (*Pinna nobilis, P. rudis, Atrina pectinata*), living on soft bottoms, suffer serious direct physical damages by human activities such as anchorage or trawling of boats (Scotti and Chemello, 2000). Attempts have been made to replant *Pinna nobilis* in the Marine Protected Area of Tavolara in Sardinia (Caronni *et al.*, 2007), following the suggestions by Attilio Cerruti (1878-1956) who tried similar experiments in the Mar Piccolo of Taranto (Cerruti, 1937; 1939) to encourage the local production of byssus (**Chapter 8**).

Figure 100: The Mediterranean date-mussel (*Lithophaga lithophaga*).

Trawling also damages many other large molluscs, especially deep-water gastropods, characterized by a slow growth rate and which, over time, have become increasingly rare, such as *Tonna galea* (Fig. **101**), *Charonia lampas* (Fig. **101**), *C.*

tritonis, and *Ranella olearia*. Already in 1918, Raffaele Issel (1878-1936), the founder of the Italian marine biology, wrote: *Unfortunately, the* Dolium (*i.e.,* Tonna) *seems to have disappeared around Genoa and Naples; common a few decades ago, it is now extremely rare*. It is often difficult to understand the reasons for a species, or a population to go into crisis, but generally it is fault of Man. The depletions in certain Mediterranean localities of the "ear of the sea" (*Haliotis tuberculata*), the giant limpets (*Patella ferruginea* (Fig. **102**)*, P. nigra*), the purple murex (*Thais haemastoma*) and *Fasciolaria lignaria* may be the result of, in many cases, excessive harvesting for food, but also by a general deterioration of the environment of the Mediterranean coasts. Although the decline of the typical organogenic platforms of the southern Mediterranean built by *Dendropoma petraeum,* an intertidal sessile gastropod, can be traced back in time, the continuous use of the coastal zone by bathers has played a significant role in its demise.

Figure 101: A living *Charonia lampas* (left). *Tonna galea* (right).

Figure 102: *Patella ferruginea*, a protected Mediterranean limpet, lives in the tidal zone of the rocky littoral.

Finally, some species are protected by strict international laws, but controls are often lacking and very often they are still harvested indiscriminately. Other species, not yet taken into account, must be protected as soon as possible (Alves, 2012). However, in the logistics of proper actions for conservation and, in general, in the fight against the loss of biodiversity, it is necessary to take actions that they will not only protect a particular species, but the whole environment where the species lives. Obviously, in this case the actions are much more complex, but it is the challenge for the future.

Fossil shells

Fossil deposits are full of shells, often of great collection interest. The protection of these deposits is generally guaranteed by national laws: in Europe, for example, all fossils are property of the states which, however, may give concessions to private individuals or consortiums for the exploitation of mineral deposits. In other countries, especially those of the circumtropical region (Fig. **103**), fossil deposits are often not very protected and this causes serious damage to a unique heritage, considering that the international trade of fossils, often illegal, is widespread and very lucrative. The establishment of parks and protected areas, and an international legislature that manages the marketing, seems today more necessary than ever, along the model of the CITES rules for the protection of the living fauna and flora. A good example is found in France in the Geological Reserve National Nature of the Haute-Provence, between Alps and Var regions. Throughout this vast territory, forty-seven municipalities have chosen to join forces to protect rocks, fossils, and landscapes to preserve a history of 300 million years ago.

Figure 103: The ammonite *Newboldigera*, Tulear, Madagascar.

CHAPTER 15

Serving Science

Abstract: Molluscs have played an important role in the history of science. They inspired models in mathematics and geometry, or were protagonists of the great discoveries in geology, biology, pharmacology, and medicine.

Archimedes, Bernoulli, Descartes, Leonardo, Dürer, among others, were inspired by molluscs in their works, mainly studying the shell coils.

The origin of the fossils has always fascinated scholars of every period, and shells were at the center of important scientific and theological disputes like the problem of the Universal Deluge. So molluscs occupy an important place in the palaeontology too.

Thanks to Lamarck, they entered in the history of the biological sciences. Describing fossil shells, he firstly said that any variation of life forms is the result of an adaptation to environmental conditions and that living species are not immutable over time: a revolutionary theory reconsidered and revised by Darwin himself. Even in the physiology and biochemistry, molluscs have always been an excellent biological model for evaluating hypotheses and perform experiments. Gastropods and squid have had a central position in several Noble Prize studies.

In the last decades, human activities determined a rise in atmospheric CO_2 which, in turn, promotes a slight, but significant acidification of both marine and freshwaters. The molluscs, builders of calcium carbonate shells, are very sensitive to the pH changes and are often used as experimental models.

Mussels are very useful to assess the levels of contamination in coastal waters, due to their ability to focus, through bioaccumulation, chemicals with which they come into contact in their continuing filtration activity.

Finally, many molluscs have the capacity to produce biologically active substances which may affect the medicine. In this field, some important successes have been achieved.

Keywords: *Aplysia, Aqua Perlata,* Archimedes, Bernoulli, biologically active molecules, *Cepaea,* Cuvier, Darwin, date-mussels, Descartes, Dürer, fossil shells, golden section, Hodgkin, imposex, Kandel, Lamarck, Leonardo, malacotherapy, mussel-watch, nudibranchs, Nobel Prize, Pozzuoli, spiral, squid, Temple of Jupiter, Universal Deluge.

Riccardo Cattaneo-Vietti, Egidio Trainito and Mauro Doneddu

If a marine organism is fat, fleshy, slow moving, and brightly colored, National Cancer Institute wants it. Such a creature must have some mechanisms for defense against predators. Usually the weapon is a chemical.

David J. Newman (1995), *National Cancer Institute, U.S.*

Molluscs have played an important role in the history of science. They inspired models in mathematics and geometry, or were protagonists of the great discoveries in geology, biology, pharmacology, and medicine.

In mathematics, the myth tells that Archimedes (287? -212 BC) described the spiral (Fig. **104**) which has his name, inspired by the shape of *Turritella communis*, a very common Mediterranean gastropod, certainly well known by the great mathematician of Syracuse. That is no small thing, if we think that his *Treatise on Spirals* allowed to John Napier (1550-1617) defining the logarithm, at the base of differential geometry. Moreover, the great Swiss mathematician Jakob Bernoulli (1654-1705) called it *spira mirabilis* and wanted it on his tombstone, with the following inscription: *Eadem mutata, resurgo* (although changed, reborn). Most probably also René Descartes (1596-1650) was inspired by the coils of *Nautilus*, a tropical shell very popular in the 17th century in Europe, when he described his logarithmic spiral (Fig. **104**), according to which the curve widens constantly, moving away from the origin, following a Fibonacci sequence (1170-1250), where each digit is given by the sum of the previous two. In fact, in molluscs, as well as in many other organisms, the logarithmic spiral is the growth model that better integrates the relations between physical developments, employment space and energy savings, according to D'Arcy Thompson (1961).

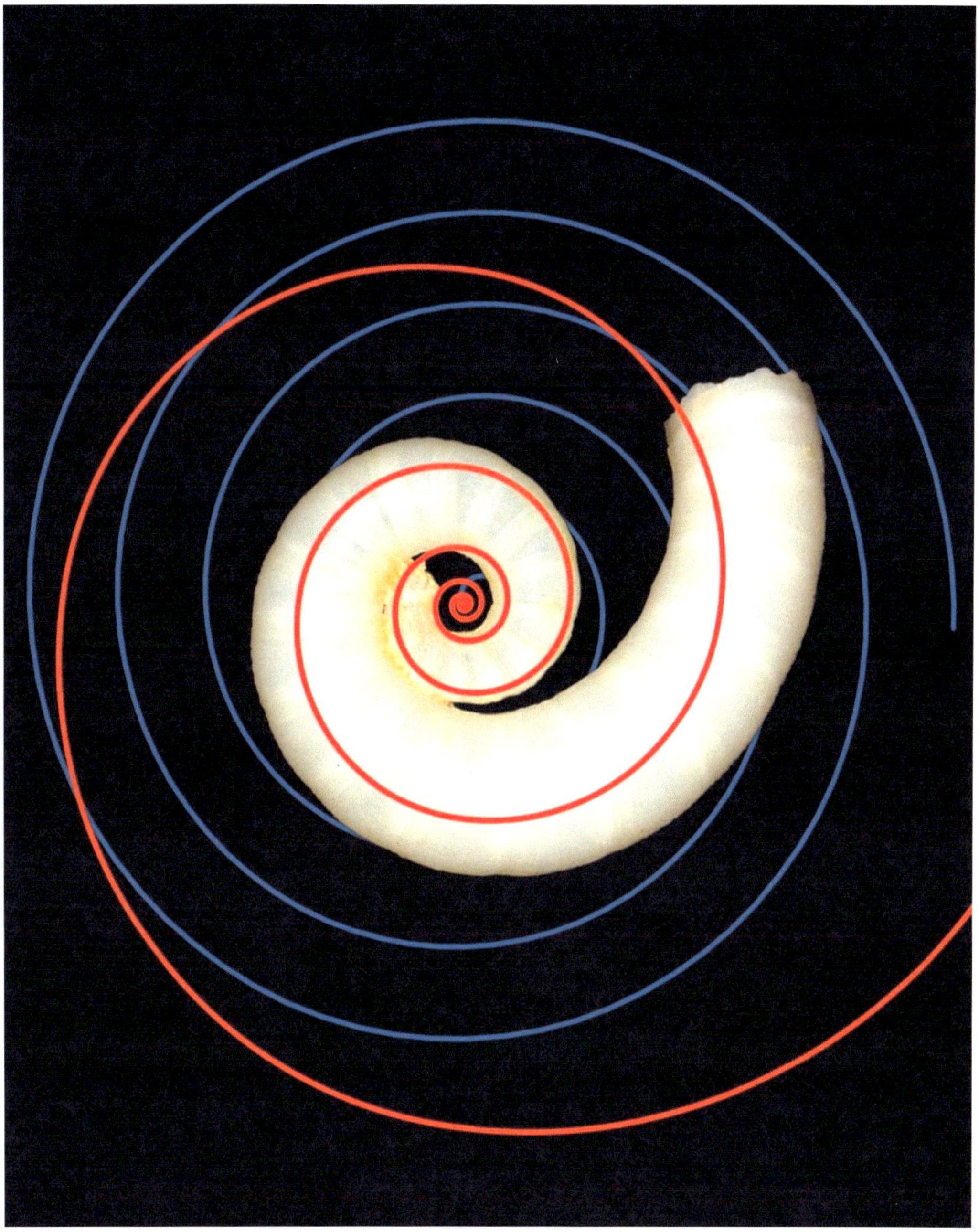

Figure 104: The Archimedes and Cartesian spirals and the *Spirula* shell.

So following the Cartesian spiral logics, the cephalopod increases the volume of its shell in proportion to its body development, contrary to what happens in the Archimedes spiral, where there is no increase, but only a constant elongation of the spiral itself.

The logarithmic spirals of many shells seem to follow a mathematical model linked to the "golden section", the so-called *Divina Proportione*, the geometric relationship between the part and the whole thing, which remains constant and harmonious. In other words, in the proportion of two magnitudes, the ratio between the lesser and the greater is equal to that existing between the greater and the sum of the two and that between the lower and the difference between the two. The "golden section" is the basic model to which many natural forms seem to obey and consequently inspired greatest sculptors, painters, and architects of all time, until to Le Corbusier (1887-1965).

In the Renaissance, many artists in fact began to take an interest in the geometry of natural forms, for understanding the Nature's theoretical models, in an attempt to interpret the natural harmony, and find the "hidden print of Creation": a task, evidently very presumptuous, but that has always fascinated artists and scientists. Also Leonardo da Vinci (1452-1519) studied the gastropod spirals, which probably inspired the projects of the helical staircases of the castles of Chambord and Blois, both along the Loire Valley. Albrecht Dürer (1471-1528), the refined German painter and engraver, followed this path. He studied a lot the shell geometry, describing, among other things, a family of three-dimensional geometric curves, the "konchoides of Dürer", which he called *muschellini*, from the German word for shell. To Dürer is also due the "spiral of Pascal," a flat algebraic curve, whose shape resembles a snail shell.

Today, thanks to computer analysis and fractal geometry, the whorls, and the complex geometric designs of the shell were studied in detail, defining mathematical models and algorithms (Meinhardt, 1995).

Fossil shells and the Universal Deluge

The origin of the fossils has always fascinated scholars of every period. In the first centuries after Christ, fossils were generally considered remains of organisms not able to survive the Universal Deluge.

Tertullian (about 155-220 AD), an important early Christian theologian and moralist, in his *De Pallio*, wrote: *There also was a time that all the Earth has changed and is covered by all the water that exists. Even today, crustaceans and circular shells from the sea, staying abroad in the mountains, to prove to Plato that even the steepest sections were flooded.*

Even in the Middle Ages there differed greatly from that interpretation: the franciscan Ristoro d'Arezzo (1223-1283) in his work, *La composizione del mondo colle sue cascioni*, the first scientific text written in vernacular, claimed to have found many "nicchi" (shells) in the mountains: a sign that the mountain was made from the Deluge ... (Mottara, 1999). The same Giovanni Boccaccio (1313-1375), in his early novel *Filocolo*, described the fossil shells from the Val d'Arno (Florence) and presupposed that the sea had covered those lands: *in ... Tuscia, in the midst of which, almost to beautiful floors, stands a small hill, which the waters, avenging the just wrath of Jupiter, when the sins of Licaon deserved to do flood the world, left second the opinion of many, which I consider true, however, that, highlights of this truth shows the small hill full of marine shells ...*

In the Middle Ages, the world view was still that of the Holy Scriptures and the Nature was a reflection of the divine plan of Creation. God created the living beings, and Adam gave them a name. The species, created by God, have always remained the same, unchanged. At best, it was possible to think that some species were migrated from another part of the Earth, but not disappeared. The hypothesis that, in later times to the Creation, some species could appear or disappear questioned the absolute perfection of the divine plan: how a species could disappear if created by God himself, and especially how a new species could suddenly appear when the Creation was already occurred? The extinction of a species was unacceptable, having been created by God himself. The same concept of Fixity, the theory that the all species were created by God, elaborated by the 18th century scientists, partially absorbed these ideas. Accepting other cases than those suggested by the Bible, scientists could easily fall into a very dangerous heresy.

Although Avicenna (980-1037), the famous Persian physician and philosopher, Augustine (AD 354-430) and Thomas Aquinas (1225-1274), one of the most influential medieval thinkers of Scholasticism and the father of the Thomistic school of theology, suggested that the fossils were only the remains of ancient organisms, having nothing to do with the Deluge, only in the Renaissance, the theory of the Universal Deluge will be slowly abandoned for explaining the presence of marine fossils in the mountains.

In fact, at the beginning of Renaissance, Fra Girolamo Castoro (1476? -1553), philosopher and physician, argued that the fossil shells could not only be the result of the Deluge, and hypothesized that they were the remains of marine organisms lived in an era covered by sea in the past. In the same years, Leonardo da Vinci (1452-1519), a great observer of Nature, suggested that the huge deposits of fossil shells, so common in Lombardy and Tuscany, were not a result of the Universal Deluge, as declared by the scholastic orthodoxy, but due to the accumulation, along the time, of the remains of once living organisms. An observation in contrast with the concept of uniqueness of the Universal Deluge: *Oh folly and sim-*

plicity of those who wish that the animals in these shells come from distant seas, carried by the flood.

Also Ferrante Imperato (1550-1631), Neapolitan pharmacist and naturalist, in his *Dell'Historia Naturale* (1599), supported this hypothesis, but even in the 17th century, the biological origin of fossils continued to encounter great difficulties of acceptance and many scholars continued to think that fossils were evidences of the Deluge or just "freaks of nature" or real "works of the devil."

However Antonio Vallisneri (1661-1730) in his work *De' corpi marini, che su' monti si trovano; della loro origine e dello stato del Mondo avanti 'l Diluvio*, expounded his theories on the origin of fossil shells that were found in the mountains, refuting the hypothesis that they were "freaks of nature", the remains of ancient meals eaten on the mountains, sea animals hatched from eggs mysteriously leavened in offshore, or other curious theories related to the Deluge. According to Vallisneri, their presence on the mountains was, however, due to the emergence of land formerly covered by the sea, following geophysical upheavals of the Earth's crust. This hypothesis was widely accepted, but still had not included such close links between the fossil and living forms. At last, the increasing knowledge on the morphology of animals and plants permitted to distinguish the rocks patterns from the biological forms.

Finally, Robert Hooke (1635-1703) described the similarities between fossil ammonites and nautilus, using one of the first microscopes: so it was increasingly difficult to argue that the fossils did not have any relationship with living organisms. Anyway, it will still be a long and bumpy process that will begin to unfold only with Lamarck and Darwin, between the end of the 18th century and the middle of the next.

The Leonardo intuition was amazing, considering that even at the end of the 17th century, Martin Lister (1638-1712), the English author of the important *Historia*

Conchyliorum (1685-1692), while recognized the similarity among living and fossil forms, considered inorganic imitations produced in the rocks, those more difficult to interpret, at least until the discovery of its analogue living. Three centuries before the birth of the modern paleontology, Leonardo had pioneered the scientific method based on observation of the natural event, the formulation of a hypothesis and its confirmation by subsequent observations. However, his writings, in reality only notes written in vernacular and ever organized in a scientific treaty, had not spread, and that they were deserved for a long time, remaining unknown, affecting very little to the scientific thinking of the time.

The molluscs, therefore, occupy an important place in the history of paleontology: we now know, for example, that the ammonites are excellent guide fossils, in particular for the Mesozoic: they are, in fact, a valuable tool in biostratigraphic evaluation, for dating sedimentary rocks to a precise geological age, and to define the relationships between sedimentary layers of different geographic regions, even on a global scale.

Even the common Mediterranean date-mussel (*Lithophaga lithophaga*) has a reserved place in the history of science, thanks to its ability to bore the limestone: the columns of the Temple of Jupiter Serapis in Pozzuoli (Naples) perforated by this bivalve, are, in fact, remembered in all texts of the geology of the world.

Another small group of bivalves, the ship-worms, have a small place in the history of the civil engineering. They are able to dig wood and are greatly feared by sailors (**Chapter 13**): the shape and the action of their shells inspired the English engineer Marc Isambard Brunel (1769-1849) who, in order to build under the Thames a tunnel in particularly unstable muddy sediments, created a shield tunneling, clearly suggested by the shell shape of this small mollusc.

The date-mussels of Temple of Jupiter Serapis in Pozzuoli

*The behavior of the Mediterranean date-mussel (*Lithophaga lithophaga*), well known in gastronomy, to bore the rock, made it famous in the history of geology. Close the port of Pozzuoli (Naples), it is indeed possible to admire the Corinthian marble columns, almost 12 m high, of the Temple of Jupiter Serapis. These columns are the remains of a market (*macellum*), probably built between 68 and 96 AD, which owes its name only to the find of a statue of Serapis, an Egyptian deity, during its excavation (Fig. **105**).*

Figure 105: A 19th century view of the Temple of Jupiter Serapis in Pozzuoli (Naples).

The temple represents one of the world's most famous examples of bradyseism, the vertical movement of the continental shelf, due to pseudo-volcanic processes. Charles Lyell (1797-1875), the founder of modern geology, devoted large space in his Principles of Geology *to these processes and made the columns of this temple and their date-mussels famous in the world. The columns show, in fact, the typical excavations due to the date-mussels boring action, about 3.5 m from the base: undoubted witness that those columns had been underwater for a certain period. In the Middle Ages (9th-10th centuries), in fact, a slow but steady bradyseism led to a* *lowering of the land of almost 6 m, and the temple remained completely submerged for centuries. Later, between the 12th and 16th centuries, with a reverse phenomenon, the ground lifted, taking the current position, a few meters above sea level. In the meantime, the columns of the temple were buried under a thick layer of sediment and protruded from the ground only in part. In the second half of the 18th century, Charles III Bourbon (1716-1788), King of Naples, understood that these outcrops could hide something and dug, thus highlighting the temple as we see today.*

Molluscs in the biological sciences

The first scientific work dedicated entirely to molluscs, *Ricreatione dell'occhio e delle mente nell'osservatione delle chiocciole, proposta a' curiosi delle opere della Natura*, by the Jesuit Filippo Bonanni (1638-1725) was published in Rome in 1681. However, pictures and descriptions, more or less fanciful, had already appeared in previous treaties. Finally, during the Illuminism, molluscs have had the attention they deserved, making an important contribution to the study of biological evolution.

Thanks to Lamarck, they entered in the history of the biological sciences. Describing fossils, and in particular the fossil shells of the Paris basin, Jean Baptiste Pierre Antoine de Monet, Chevalier de Lamarck (1744-1829), author of *Système des animaux sans vertebres* (1801), and especially *Philosophie Zoologique* (1809), firstly said that any variation of life forms is the result of an adaptation to environmental conditions and that living species are not immutable over time, as claimed by the fixists, but in a continuous and gradual evolution. These revolutionary ideas will struggle to spread in the early 19th century in Europe where the Fixity Theory was dominant. In France, Lamarck found a formidable opponent in Georges Léopold Chrétien Frédéric Dagobert, Baron de Cuvier (1769-1832), famous fixist and founder of comparative anatomy and paleontology. Cuvier was the proponent of the *Theory of natural disasters*, according to which the fossils were nothing more than the remains of extinct species as a result of catastrophic events, a theory opposed to the idea of gradual evolution of Lamarck and that certainly raised less scandal, referring, in part, to the Universal Deluge. On the other hand, at the beginning of the 19th century, the Church's opposition, the French Revolution fear, and the following Napoleonic Wars have certainly not encouraged the spread of the thought of a French researcher in Britain and Germany, and for many years the evolutionary ideas of Lamarck were refused. Curiously, own an Englishman, Charles Robert Darwin (1809-1882), will reconsider and revise, fifty years after, the *Theory on the inheritance of acquired characteristics* by Lamarck, allowing speculations which will be at the basis of his *Theory of Evolution*. With *On the origin of species*, defining the role of the natural selection in favoring the best adapted organisms along the struggle for life, published in London in 1859, Darwin made one of the major cultural revolutions of all time.

Even today, in macroevolution studies, the bivalve fossils are crucial to understanding the processes of mass extinction and radiative speciation, the basic concepts of the *Theory of punctuated equilibrium* (1972) by Niles Eldredge and Stephen Jay Gould in in which is assumed that species, in their evolutionary histories, tend to remain stable for long periods of time, while the speciation processes occur in a relatively short time.

Even in the ecological studies, molluscs have done their part: the concept of biocenosis, the association of organisms that live permanently in a given habitat, was applied for the first time in 1877 by the German zoologist Karl August Möbius (1825-1908) for describing the structure of the oyster beds of the North Sea. Again, the concept of ecological niche, the theoretical species-specific multidimensional space which sums up all the vital needs of a species, has been partially understood by studying the trophic preferences of *Hydrobia*, a group of small gastropods common in Scandinavian estuarine environments. Some species belonging to this genus, very similar, and living in the same habitat, are in fact able to slightly change their trophic needs to decrease the interspecific competition (Fenchel, 1975).

In the genetic population study on the relationship between predation and genetic changes, an important role has taken a European land snail, *Cepaea nemoralis*, a species showing highly polymorphic colours of its shell.

Figure 106: The different colour patterns in the European land snail *Cepaea nemoralis*.

This polymorphism (Fig. **106**) can be expressed in yellow, pink, or brown and is generally controlled by the thrush, the main bird predator of this snail (Pianka, 1994). The colour of the shell is, in fact, a mimetic character which decreases the pressure of the bird's predation, and consequently darker shells tend to be more common in the underwood habitats, while in the sunniest meadows, snails show lighter colours, such as yellow or multichromatic bands. In addition, the brown shells are generally more common in North than in South Europe, because in colder regions the darker shell warms faster at sun, allowing the snails to be more active than those of light colour. In 2009, a European project (*Evolution Megalab*) assessed how, over the past 30 years, the geographical distribution of the different colour patterns changed in Europe, suggesting that the global warming could affect the distribution of the genetic characters of a species on a large scale.

Moreover, in genetic studies, the manipulation of pond snails (*Lymnaea*) embryos in laboratory, allowed to alter the spiral shells of these snails, demonstrating the importance of the embryo structure in determining the physical appearance.

In the last decades, human activities, in particular industry and farming, also determined a rise in atmospheric CO_2 which, in turn, promotes a slight, but significant acidification of both marine and freshwaters. The molluscs, builders of calcium carbonate shells, are very sensitive to the pH changes and are often used as experimental models.

The biological effects of the CO_2 increase dissolved in sea are studied also thanks to the morphological and structural changes in the shells of many molluscs (pteropods and limpets, for example), as a result of acidification itself, which can reduced or altered significantly the content of calcium carbonate (Hall-Spencer *et al.*, 2008). The very long-lived bivalve shells can also be very useful to obtain information on the temporal temperature in the water masses of a certain geographical area: Strom *et al.* (2004), analyzing the structure of the shell geochemistry of the "geoduck" *Panopea generosa*, a very long-living bivalve, were able to reconstruct the trend of the temperature of the sea water in the Strait of Juan de Fuca (State of Washington) since 1877, showing that the 90s years of the 20th century were, in that region, the warmest decade in the last 120 years.

Another species that could be taken into account when assessing the effects of global warming in the Mediterranean is, for example, a typical small gastropod, *Echinolittorina punctata*, a thermophilic littoral species which is expanding its distribution range in the north-western basin Mediterranean thanks, probably, to an increase in the surface temperature of the coastal water masses (Albano *et al*., 2010).

The molluscs are also a model in ethology, the science of the animal behavior. In this field, an important role was played by the octopus, *Octopus vulgaris*, considered the most intelligent known invertebrate: the octopus is, in fact, able to learn for "association and imitation," or for "look and learn" behavior, a behavior generally shown only by some mammals. For example: giving to an octopus a good prey locked inside a transparent jar, it proved to be capable of learning by imitation opening the jar for reaching the prey, once someone showed it the technique to open the jar itself.

Even in the physiology and biochemistry, molluscs have always been an excellent biological model for evaluating hypotheses, perform experiments, and deepen the knowledge of the inner workings of some biological systems. At the end of the 80s of the last century, scientific research pointed out how some gastropod molluscs (*Aplysia californica*, *Charonia lampas* and *Turbo cornutus*) were a very interesting source of some hydrolytic enzymes, glycosidases. These molecules showed a specific and selective ability of hydrolysis of the glycosidic bonds in the oligosaccharides and can be used in the structural analysis of these macromolecules certainly involved in several cellular mechanisms of extreme interest, at the basis for serious diseases such as cancer or the common viral infections.

The recent proposals to avoid the use of animals in scientific tests also include molluscs in the list of organisms to be protected. This will greatly reduce the sacrifice that these animals have paid to the scientific development and the knowledge of Man, but at the same time, will limit seriously the research.

Molluscs and the Nobel Prize

Squids and *Aplysia* species (Fig. **107**), these last sea slugs of large size, have had a central position in the biological sciences. The study of the squid nervous system has given a fundamental contribution to the neurology research, allowing to understand the electrophysiological mechanisms of the impulse transmission in the central nervous system, while, thanks to *Aplysia*, the physiological mechanisms of the cellular memory, the basis for the fight against memory loss and some dementias, were studied.

The characteristics of the nervous system of the squid have offered an amazing tool in the field of neurology. The axon of its neurons has, in fact, a diameter extraordinarily large (0.5-1 mm), such as to be able to insert microelectrodes suitable to measure changes in the electric potential inside and outside of the cell.

Alan Lloyd Hodgkin (1914-1998), Andrew Huxley (1917-2012) and John Carew Eccles (1903-1997), the Nobel Prizes for Medicine in 1963, demonstrated that the action potential of the squid giant axon is due to an abrupt and reversible change of flux of Na^+ ions, crossing the plasma membrane. In other words, they discovered the depolarization of the cell, the reversal of the membrane potential and their relationships with the external ionic concentrations of sodium and potassium. Thus, the hypothesis according to which the sodium depolarization causes a rapid rise followed by a slower descent, in the permeability of Na^+ and a slow rise in the permeability of K^+ was born. The depolarization of the membrane opens Na^+ channels on the same, allowing increased entry of Na^+ into the cell and thus determining the ascending phase of the action potential; the descending phase of the action potential is due, on the contrary, the closure of the Na^+ channels, which reduces the incoming flow of this ion, and the opening of K^+ channels, which leads to increased leakage of K^+ from the cell.

The electrophysiological analysis of brain tissue showed how the elements deputies to the transmission and processing of information are neurons, and how they, although they exist in many types, shared some salient features. In particular, the dynamic behaviors of the neuron consisting of issuing and transmission of electrical impulses stereotyped (the so-called action potentials) to other cells. The onset of action potentials is in turn induced by the reception, through the synapses, other action potentials, from neurons connected to the one under examination. One can therefore say that the action potentials represent a basic element of the code used for the transmission of information at the neuronal level.

The squid contribution in these studies was so important that Alan Hodgkin said: *It's the squid, they really ought to give the Nobel Prize to ...*

In 2000, Arvid Carlsson (1923-) from the University of Göteborg, Paul Greengard (1925-) from the Rockfeller University of New York, and Eric R. Kandel (1929-) from the Columbia University (New York) won the Nobel Prize in Medicine for their discoveries concerning the signal transduction in the nervous system. In particular, Kandel devoted his entire life studying the nervous system of a large opistobranch, *Aplysia californica*. He demonstrated that the memory storage is in the synapses and that it is possible, thanks to the plasticity of the synapses themselves, changing their efficiency (Kandel, 1979): in other words, learning and memory appear related to the change of the synaptic function. In particular, the phosphorylation of proteins in the synapses plays an important role in the generation of the short-term memory, while for the development of a long-term memory, it is necessary to also a change in the synthesis of proteins, which may lead to alterations in shape and function of the synapse itself. These studies are the basis of our understanding of the complex processes of storage of the nervous system and also allowed us to develop new types of drugs to improve memory function in patients with different forms of dementia.

Figure 107: *Aplysia punctata.* The physiological mechanisms of the cellular memory were studied for the first time in the aplysids.

Moreover, the research on the ubiquitin, a small protein at the base of protein degradation, were led by the biochemist Israeli Avram Hershko (1937-) who received the Nobel Prize for Chemistry in 2004 with Aaron Ciechanover (1947-) and Irwin Rose (1926-2015), also working on fertilized eggs of clams. Today, we know that this biochemical process is universal, common to all eukaryotic organisms: the ubiquitin, in fact, regulates the cyclins, proteins at the base of cell division. This discovery opened new unexpected ways in the fight against the development of cancer and neurological and muscular diseases.

In applied ecology and environmental monitoring

The protection of the natural environments requires a continuous action of analysis and control: the presence or better the disappearance of some species is certainly an indication of a change in water quality, and the use of some molluscs as biological indicators in a specific environmental condition has now entered the practice of many control programs worldwide.

The physic-chemical analysis show, in fact, any alteration on time, while the structure and dynamics of the biological community are the result of the "history" of that environment, of all events that have taken place, changes very hard to detect using methods of chemical analysis.

In Europe, the quality of inland waters is assessed for over 30 years using an index, EBI (*Extended Biotic Index*), that using the presence/absence of certain species and knowing their ecological needs, allows to classify the water quality of a river.

All water basins are characterized by the presence of more or less sensitive species to pollution: consequently only the more “robust” can survive to a significant change. The increase of nitrates concentration in waters favors, for example, a major development of macroalgae which, in turn determines an increase in number of herbivorous gastropods: certainly a sign of an environment shift.

Today, European directives tend, however, to encourage not only the application of biotic indices, but also comparative analysis with blank sites for evaluating possible alterations.

Regarding the marine coast, the Framework Directive of the European Community (WFD 2000/60/EU) introduced an innovative approach for the assessment of the ecological quality of the coastal water in Europe, also studying the composition of the hard bottom benthic communities: in this sense, the presence of mussels, known to be indicators of a high concentration of organic matter in the waters, is certainly an index of low ecological quality.

Finally a small Mediterranean nassarid, *Cyclope neritea*, has to be mentioned as it seems to bear well anoxia and toxicity in the hydrothermal waters of Southern Tyrrhenian Sea and Aegean Sea, thus becoming a kind of biological indicator of these very special environments, characterized by an abnormal development of sulphobacteria (Southward *et al.*, 1997).

In applied ecology, the use of strong herbivore grazers, as *Patella*, has been attempted to control the development of biofouling on the hulls of small boats, but results are still conflicting and difficult to interpret (Safriel *et al.,* 1993). Most normal and common is the use of herbivorous gastropods to control the growth of algae on the walls of a home aquarium as well as on the cages for aquaculture. Finding a biological system that effectively prevents or allows decreasing the use of highly toxic biocide in the storage of yachting now appears strategic and research of biologically active substances is very committed in this direction: some terpenes extracted from some nudibranchs seem to show a good antifouling activity in the test. These molecules appear to inhibit the settlement of barnacle larvae, one of the main components of the biofouling.

The Mussel-Watch

Mussels are very useful to assess the levels of contamination in coastal waters, due to their ability to focus, through bioaccumulation, chemicals with which they come into contact with in their continuing effort to filtration. Designed in the mid-60s of last century in the United States, the Mussel-Watch is an international project which, using mussels (*Mytilus galloprovincialis, M. edulis*) as "body guards", allows a cheap assessment of the chemical contaminants of industrial origin, such as total hydrocarbons, polycyclic aromatic hydrocarbons (PAHs), polychlorinated biphenyls (PCBs), heavy metals, radionuclides or substances coming from agriculture, such as pesticides. By placing "healthy" mussels for at least 12 weeks in a "suspicious" site and subsequently analyzing the concentration of contaminants reached in its tissue, it is possible to determine the level of contamination of that environment and provide an administrative tool for a proper management of the coastal waters (Andral *et al.,* 2004).

In marine virology, the use of filter-feeding bivalves is today indispensable too: in fact, they are able to concentrate pathogenic viruses. Asahina *et al.* (2009) have shown that some bivalves, such the tree oyster, *Isognomon* sp., can act as body guards against the possible presence of norovirus, among the most common agents of acute gastroenteritis in coastal waters. Growing on the mangroves, these bivalves are, in fact, able to accumulate these viruses and consequently, to facilitate monitoring.

A "clever" mussel

Also as part of the environmental monitoring, an interesting use of mussels has been applied to control the quality of coastal waters, especially in the harbors.
In case of stress, the mussel inevitably closes the valves. So, by applying two electrodes to the valves themselves, in case of closure of the valves a circuit is closed, permitting to understand in real time that the chemical-physical characteristics of the environment were changed negatively. Knowing the opening and closing

rhythms of the valves in controlled environments, it is possible therefore become aware of a fault occurred in the water quality.

Finally, in the case of chronic pollution of the water that could cause chromosomal alteration in organisms exposed to potentially carcinogenic substances, such as some hydrocarbons, oysters have long been considered bioindicators of great interest. Their ability to accumulate toxic substances in the tissues, allows to analyze not only the rates of bioaccumulation, but also the direct effects of these substances on their metabolism through biochemical tests that highlight possible conditions of stress or damage to DNA.

In the terrestrial environment, however, the snails are often used as bioindicators of air quality maintained for a certain period in polluted sites or may not give an idea of the very significant level of air pollution in the area.

The imposex in muricids

Gastropod imposex or pseudohermaphroditism is a pathological character used as index to evaluate the presence of toxic substances in the coastal marine waters. Imposex is, in fact, a malformation of the female genitals in a gastropod family (muricids) that, under particular environmental conditions, develops a gradual masculinization of females who become so sterile.

This phenomenon is due to continued stress conditions and in particular to the presence of toxic organic substances, such as organic-tin (tin-tributyltin, TBT), compounds introduced in the coastal waters as a result of their use as components of anti-fouling paints, before their marketing was limited (Mann *et al.*, 2006; Garaventa *et al.*, 2007).

Signs of imposex in varying degrees of severity were observed in the muricid *Hexaplex trunculus* collected in all the Italian Marine Protected Areas (Terlizzi *et al.*, 2004). This result has questioned whether there is a direct relationship between the imposex development in muricids and the presence of TBT in the environment. However, as a precaution, the use of TBT-based paints for the antifouling purpose for boats of less than 25 m was banned in UK in 1987 (Food Environment Protection Act, 1985) and widely extended in 2008 by the IMO, the International Maritime Organization. Today, its use is prohibited on any vessel wishing to enter a port of the European Union.

Biologically active substances

The pharmacology, the science that studies how chemicals interact with the living, it is always in search of biologically active natural substances that may affect the medicine (Bankendorff, 2010). Such as many other groups of animals and plants, molluscs are able to synthesize, for offensive or defensive purposes, various organic substances, often toxic, whose pharmacological activity may have an important therapeutic fall, often in the field of cancer. It is therefore easy to understand because the biopharmaceutical industry invests heavily in the research on these substances, even if it takes years before being able to isolate and test a new molecule: and often the practical results are much lower than expected. But some important successes have been achieved.

The conotoxins produced by cones, due to their low molecular weight, which made them easy to synthesis (Haddad *et al.*, 2006), are now widely used in neurology: probably the African magic medicine, which used excerpts of gastropods to cure mental diseases, had already guessed their properties. Among these conotoxins, the activity of ziconotide is very interesting, as this peptide, composed of 25 aminoacids derived from the toxin of *Conus magus*, shows excellent analgesic properties that can selectively block N-type calcium channels, a major contributor to the modulation of pain at the spinal cord level. Ziconotide has analgesic efficacy comparable or superior to that of morphine.

Another group of peptides of pharmacological interest has been isolated in *Dolabella auricularia*, a large Indian Ocean gastropod: its dolastatines seem to give promising results in the fight against liver and breast cancers, and in cases of leukemia (Pettit *et al.* 1997; Poncet, 1999), while from the Japanese *Aplysia kurodai*, a good cytotoxic, the aplironine A, has been extracted. Encouraging extracts (called paolin, paolin I and II) from *Haliotis* seem to have antibacterial and antiviral activity, while the mercenine extracted from the "quahog", *Mercenaria mercenaria* (Fig. **46**), would act as an inhibitor in some types of cancer in mice. Important against prostate cancer are also substances present in the blood of *Concholepas concholepas* (Fig. **109**), while the freeze-dried extract of *Perna canaliculus*, the New Zealand green mussel, seems to have a high anti-inflammatory action: its eicosatetraenoic acid (ETA) has a strong ability to reduce the inflammatory response and is useful in the treatment of rheumatoid and arthritic diseases. Its extract, marketed since 1986 as *Greenhealth Green Lipped Mussel*, is patented.

One of the richest sources to detect biologically active substances seems to be the nudibranchs on which at least 50 years has focused the interest of researchers (Avila, 1995). Lacking the protection of the shell, these gastropods have developed the ability to metabolize various toxic substances or repellents, often terpenes: these chemical weapons can also be of food origin, present in their prey, and that the slugs are able to accumulate in their tissues or in special glands for use in its defense. Some species can operate slight modifications of these molecules, increasing or decreasing the toxicity, other are, on the contrary, capable of independently produce toxic substances, regardless of their diet.

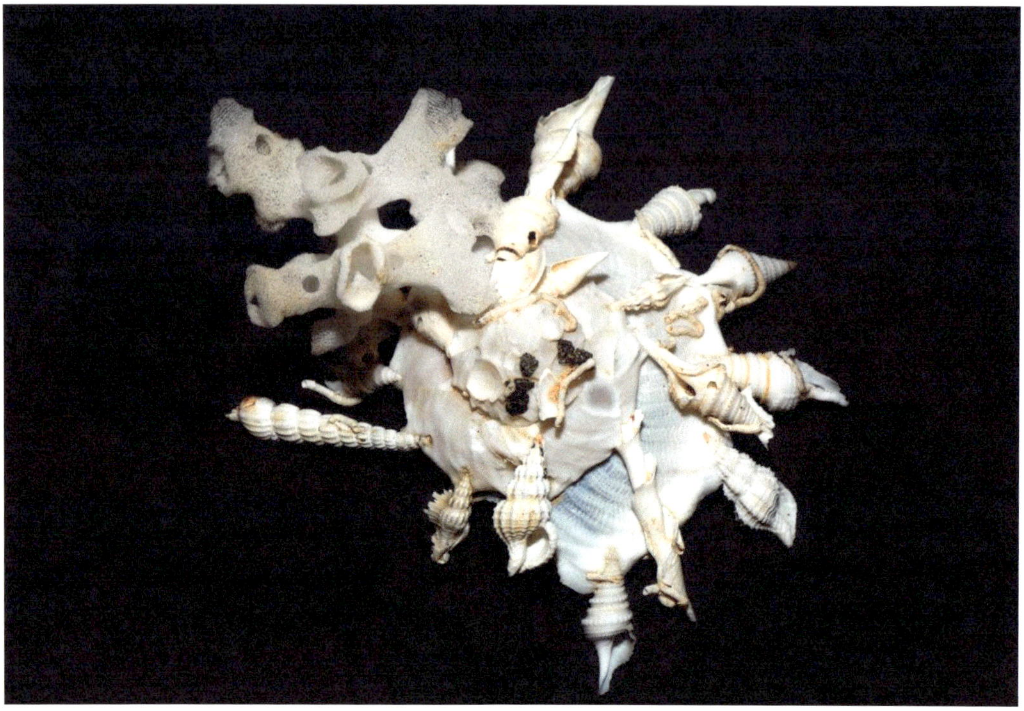

Figure 108: *Xenophora pallidula*, a gastropod which can "paste" small objects to its shell: these adhesives could have a future in surgery.

Natural substances that seem to have an interesting in different industrial sectors are the components of the cement of Xenophoridae (Fig. **108**), a group of gastropods, called carrier-shell, which can "paste" small objects to their shell: these substances could have a future as adhesives in surgery. Also the adhesive ability of the byssus on very smooth surfaces such as glass has been considered: its adhesive properties could, in fact, suggest new biomaterials with unique characteristics in terms of mechanical strength. The delicate nature of these studies suggests an obvious degree of secrecy, and only years after their discovery, the effectiveness of these substances will be public.

The nudibranchs: a source of new biologically active molecules?

Nudibranchs are gastropods without shells. They are gorgeous and still poorly understood. For colour and size, reaching a maximum of fifteen centimeters, are called "sea butterflies". Along the Mediterranean coast, more than two hundred species are known, but it is in the tropical seas that they have reached an extraordinary development: in fact, only the Great Barrier Reef is home to over 2,000 species.

The sea slugs, as well as being incredibly attractive for their beauty, have the capacity to produce biologically active substances, and have been for years "in the target" of the pharmacological research. Having, in their evolutionary history, removed the shell, they developed to protect themselves the ability to produce or store toxic substances (often coming from the diet, mainly cnidarians, bryozoans, and sponges). Even their colour, often very lively, is a signal that warns potential predators that they are "toxic or indigestible." The science, always looking for chemicals of pharmaceutical interest, studies them since many years to isolate useful substances.

Figure 109: *Concholepas concholepas* shell from Chile.

The malacotherapy

Today, the ethno-medicine is studied to understand the substances, traditionally used in the past, and their effect on the human health (Alves & Souto, 2015). In general, we think of the ethnic medicine refers to non-European cultures, but in reality, even the western medical-pharmaceutical system, born in Europe thanks to the Medical School of Salerno between 10th and 13th centuries, did nothing that to study, just as the traditional Indian and Chinese medicine, the curative products present in many plants and animals, following an absolutely deductive logics. In Salerno, in fact, for a lucky set of circumstances, the greatest scholars of the time were able to rediscover the medical knowledge of Arab and Greek-Byzantine cultures, almost lost in Europe in the Middle Ages.

Among the species that have provided and continue to provide biologically active substances and useful to Man, many molluscs (Alves *et al.*, 2013; Alves & Rosa, 2013).

The first knowledge of their healing powers belongs to the traditional pharmacopoeia as Hippocrates of Kos (460-377 BC) already said. The Father of Medicine claimed that raw oysters were laxative and that their cooking broth can cause constipation and flatulence.

Take five African snails, roast them, crush and dissolve in myrtle wine with a little of acacia: this will solve your dysentery problems, Pliny suggested 2,000 years ago (Saul, 1974). Pliny was known to be a sucker, writing down any information without a minimum critical analysis, so many of his statements are fanciful. In this case, it is likely that it was the calcium carbonate of the shell to solve the problem, because a broth of mussels is the perfect solution to move the intestine, in short, it has a light laxative action. Al contrary, the ashes of pulverized shell dissolved in honey were a good remedy for wounds and bites: still the anti-hemorrhagic action of dust shells is known in the traditional African medicine.

Another Father of classical medicine, Galen (129-216 AD), held that the broth and the ashes of snails were shown to cure the evil eye, hernias, ulcers and to combat the fever (Voultsiadou, 2010; Voultsiadou *et al*., 2010), while the shell ashes, according to Dioscorides Pedanius (40? -90? AD), were also very useful in the cleaning of children's teeth and maintain healthy the gums: still now shell dust is added in many types of toothpastes as their slightly abrasive action. The same rules that belong to the Roman medicine, taking mostly from the Greek tradition, they suggested to use both molluscs that their shell to cure stomach pain, headache, tooth-ache. Even the medieval apothecaries were very fond of the belemnites fossils, which they attributed, special powers to cure nightmares and spells, heal wounds, and clean teeth. Even today, the Andean people use pearl powder to treat wounds, while the Vietnamese traditional medicine suggests shell powder (i.e., calcium carbonate) to treat stomach acid or stopping bleeding.

Land snails were used to cure the diseases of liver, the thinness, and certain cases of wasting: the personal physician of Pope Innocent X (1574-1655) advised snail as food for convalescents who need to regain the lost appetite and vigor.

In Europe, between 1600 and 1800, was also widely believed that the oyster possesses therapeutic properties to treat tuberculosis, pulmonary catarrh, and dysentery, and above all, that it possessed aphrodisiac qualities. In fact, in the aphrodisiac properties of the oyster is something true: they contain, in fact, a good amount of zinc, an element that favors the synthesis of testosterone in humans.

Still in '700, when Karl von Linneus (1707-1778) described the Mediterranean cuttlefish, *Sepia officinalis*, called it the "curative" cuttlefish. He knew that, according to the classic pharmacopoeia, a lotion of *Sepia* was a good remedy against asthma and bronchitis and suitable for combat depression during the menstrual cycle, pregnancy and menopause. According to Arab tradition, the cuttlefish bone, pulverized and mixed with honey and passed on the eyes, heals the white spots of the eye (Scarpa, 1980). Moreover, wraps made with abalones (*Haliotis*) are recommended in traditional Chinese medicine to treat visual disturbances (Guerci, pers. com.).

In *Encyclopedic Dictionary of Medical Sciences* (1866) recipes to clear the bronchi are still listed for preparing oily syrups with extracts from land snails or winkles, the common small gastropods along the Atlantic coasts of Europe. In England, in the 19th century, it was told that a diet of live snails or slugs cures anemic persons and sickly children. Snails were prescribed by local physicians for a number of complaints, including asthma, dropsy, eye troubles, rheumatism, and corns. Slime of slugs and snails were counted a sure cure for eczema (Rogers, 1908). Today, the pharmaceutical industry places on the market, with considerable success, syrup-based extract of snail which play a soothing cough, whooping cough and chronic bronchitis, while in China is an ancient practice to use the slime of snails, rubbed on the wrists, to bring down the fever (Guerci, pers. com.).

The elicine, a portentous snail cream

In 1980, Fernando Bascuñan, a Chilean snail farmer, observed that his employees, in constant contact with the snails, had surprisingly soft hands and that any wounds quickly healed, without conspicuous scarring or infection. Analyzing the mucus of the snail, he found that a cocktail of different substances that exerted on the skin acted as a natural exfoliating substance, thanks the glycolic acid, with strong nourishing and antioxidant properties (thanks to a protein-vitamin complex). The product improved firmness and elasticity to tissues, due to the collagen and the elastin and, more importantly, thanks to the allantoin, a protein with an excellent regenerating action. After 15 years of research, Fernando Bascuñan produced a cream, the elicine, today an international patent, marketed in over 30 countries.

Pearls, possible source of health and fitness?

People have attributed to pearls important symbolic meanings, associated also to important healing properties. According to the Arab and Indian pharmacopoeia, in fact, the pearl powder, mixed with honey or wine, strengthens the view and develops a good anti-inflammatory activity. In Europe, their supposed healing properties will be resumed in the Middle Ages: the same Albert the Great (1206-1280), Doctor of the Church and alchemist, suggested the use of powdered pearls to treat mental disorders, "love sickness", hemorrhages and dysentery.

Throughout the Middle Ages, the importance of the healing pearl will be much emphasized. Still at the beginning of the 17th century, Anselmus de Boodt (1550-1632), a student of mineralogy as well as court physician of Rudolf II of Habsburg (1552-1612), suggested to recover vigor to use the *Aqua Perlata*, a drink made from crushed pearls dissolved in vinegar and lemon, added with rose water, strawberry juice and other plants (Taburiaux, 1985). A rather expensive recipe, but the Emperor had a great passion for alchemy and the occult sciences, and the Court of Prague was attended by a host of scientists, alchemists and even some charlatan. It is easy to think that he might taste of everything and no expense spared.

The "pearl-health" combination was widespread around the world: the Egyptian physician Abu al-Abbas Ahmad ad-Dahmasni, in 1773, wrote a treatise, claiming they do stop the disease and preserve human health (Scarpa, 1980).

Today, according to the "New Age Medicine", a heterogeneous set of practices with often esoteric connections, beads with stones and precious metals constitute powerful remedies against diseases: the “precious-therapy” is still in vogue in Madagascar, in Latin America (particularly in Brazil), and in Buddhist monasteries of Tibet, where even today pills with precious metals and stones are produced for various diseases (Guerci, pers. com.). In Mexico, even in the 20th century, mother-of-pearl from unionids dissolved in lemon and other substances, is used to produce a kind *of Aqua Perlata* that leverages the scars and makes the skin beautiful, while pulverized shells were still used to stop minor bleeding (Scarpa, 1980).

REFERENCES

Albano, PG, Sabelli, B, Adani, M & Pinardi, N, (2010) The termophilous species *Echinolittorina punctata* as a new descriptor of tropicalization in the Mediterranean Sea. First data. *Biologia Marina Mediterranea,* 17 (1), 90-93.

Alves, RRN, Rosa, IL, Léo Neto, NA & Voeks, R (2012) Animals for the Gods: Magical and Religious Faunal Use and Trade in Brazil. *Human Ecology*, 40 (5), 751-780.

Alves, RRN, Oliveira, TPR, Rosa, IL & Cunningham, AB (2013) Marine Invertebrates in Traditional Medicines. In: Alves RRN, Rosa IL (Eds) *Animals in Traditional Folk Medicine: Implications for Conservation*. Springer, Berlin. 263-287.

Riccardo Cattaneo-Vietti, Egidio Trainito and Mauro Doneddu

Alves RRN & Rosa IL (2013) *Animals in Traditional Folk Medicine: Implications for Conservation.* vol 1. Springer-Verlag, Berlin Heidelberg.

Alves, RRN & Souto, WMS (2015) Ethnozoology: A Brief Introduction. *Ethnobiology and Conservation*, 4 (1), 1-13.

Andral, B, Stanisiere, JY, Sauzade, D, Damier, E, Thebault, H, Galgani, F & Boissery, P (2004) Monitoring chemical contamination level in the Mediterranean based on the use of mussel caging. *Marine Pollution Bulletin*, 49, 704-712.

Andrews, CAE (1981) *Catalogue of Egyptian antiquities in the British Museum. Jewellery 1. From the earliest times to the Seventheenth Dynasty*. 6, London.

Angeletti, S & Ferrario, M (1983) *La più grande enciclopedia delle conchiglie di tutto il mondo*. A. Peruzzo Ed, Milan. 1-792.

Asahina AY, Lu, Y, Wu, C, Fujioka, RS & Loh, PC (2009) Potential biosentinels of human waste in marine coastal waters, bioaccumulation of human noroviruses and enteroviruses from sewage-polluted waters by indigenous mollusks. *Journal Virology Methods,* 58 (1-2), 46-50.

Avila, C (1995) Natural products of opisthobranch molluscs, a biological review. *Oceanography and Marine Biology, an Annual Review*, 33, 487-559.

Bailey, J (2001) *The white divers of Broome*. Pan Macmillan, Australia. 1-301.

Baker, P, Fajans, JS, Arnold, SW, Ingrao DA, Marelli, DC & Baker, SM (2007) Range and dispersal of a tropical marine invader, the green mussel, *Perna viridis*, in subtropical waters of the southeastern United States. *Journal of Shellfish Research*, 26, 345-355.

Barker, GM (2002) *Molluscs as crop pests*. CABI Publ., New York.

Battiato, A (2012a) Identificazione degli organismi acquatici del *De Re Coquinaria* di Apicio. *Bollettino Accademia Gioenia Scienze Naturali*, 45 (374), 1-9.

Battiato, A (2012b) Contributo all'identificazione degli organismi acquatici riportati nel libro IX del *Naturalis Historia* di Plinio il Vecchio. *Bollettino Accademia Gioenia Scienze Naturali*, 45 (374), 10-60.

Bax, N, Hayes, K, Marshall, A, Parry, D & Thresher, R (2002) Man-made Marinas as sheltered islands for alien marine organisms, establishment and eradication of an alien invasive marine species. In: Veitch, CR & Clout, MN (Eds), *Turning the Tide, the eradication of invasive species*. IUCN, Invasive Species Specialist Group, occasional paper 27, 26-39.

Benkendorff, K (2010) Molluscan biological and chemical diversity: secondary metabolites and medicinal resources produced by marine molluscs. *Biological Reviews,* 85, 757-775.

Bernardini, E (1977) *La Preistoria in Liguria*. Sagep Ed, Genoa, 1-231.

Berrin, K & Larco Museum (1997) *The Spirit of Ancient Peru, Treasures from the Museo Arquelògico Rafael Larco Herrera.* Thames and Hudson Ed, New York.

Bertsch, H & Marlett, CM (2011) The Seris, the Sun and the Slugs: cultural and natural history of *Berthellina ilisima* and other opisthobranchs in the Central Sea of Cortez. *Thalassas*, 27, 9-21.

Bird, S (2010) *Memories of the Cornish Fishing Industry*. Countryside Books, Newbury, 1-128.

Blanchard, M (1997) Spread of the slipper limpet *Crepidula fornicata* (L. 1758) in Europe. Current state and consequences. *Scientia Marina*, 61 (2), 109-118.

Boltovskoy, D, Correa, N, Cataldo, D & Sylvester, F (2006) Dispersion and ecological impact of the invasive freshwater bivalve *Limnoperna fortunei* in the Río de la Plata watershed and beyond. *Biological Invasions*, 8 (4), 947-963.

Bombace, G & Lucchetti, A (2011) *Elementi di biologia della pesca.* Edagricole, Milan. 1-383.

Bouchet, P, Darigal, F & Huyghens, C (1979) *Living Seashells, Molluscs of the English Channel and Atlantic coasts*. Blandford Press, Poole, Dorset. 1-144.

Bower, SM, McGladdery, SE & Price, IM (1994) Synopsis of infectious diseases and parasites of commercially exploited shellfish. *Annual Revue Fish Diseases,* 4, 1-199.

Brown, DS (1994) *Freshwater Snails of Africa and Their Medical Importance*. CRC Press. 1-305.

Buhle, ER & Ruesink, JL (2009) Impacts of invasive oyster drills on Olympia oyster recovery in Willapa Bay, Washington, United States. *Journal Shellfisheries Research,* 28(1), 107-112.

Bulgarelli, F & Melli, P (2001) L'abbigliamento del pellegrino. In: Bulgarelli, F, Gardini, A & Melli, P (Eds), *Archeologia dei pellegrinaggi in Liguria*. M. Sabatelli Ed., Savona. 1-160.

Bussani, M (1983) *Guida pratica di mitilicoltura*. Ed. Edagricole, Bologna.

Buzzati, D & Cinotti, M (1966) *L'opera completa di Bosch*. Rizzoli Ed.

Caddy, JF & Rodhouse, PG (1998) Cephalopod and groundfish landings, evidence for ecological change in global fisheries? *Revue Fish Biology and Fisheries,* 8, 431-444.

Caprotti, E (1979) *Simbologie sacre e conchiglie. 1. Le culture primitive e l'Occidente*. New Press, Como. 1-48.

Caprotti, E (1981) Simbologie sacre e conchiglie. 2. Le culture tradizionali dell'Oriente. *Bollettino Malacologico, Milan*, 17, 229-244.

Caprotti, E (1985) Per una storia della malacologia dall'età barocca al secolo dei lumi. *Bollettino Malacologico, Milan*, 21, 183-202.

Caprotti, E (2009) La metafora dell'ostrica e la condizione umana. *Bollettino Malacologico, Milan*, 45, 99-104.

Caprotti, E (2010) Antiporte malacologiche del Settecento. *Bollettino Malacologico, Milan*, 46, 16-28.

Carazzi, D (1893) *Ostricoltura e mitilicoltura*. Hoepli Ed., Milan.

Cardini, F (1988) La conchiglia e la perla. *Abstracta,* 26, 46-53.

Carefoot, TH (1987) *Aplysia*, its biology, and ecology. *Oceanography and Marine Biology, An Annual Review*, 25, 167-284.

Carlton, JT & Geller, JB (1993) Ecological roulette, the global transport of nonindigenous marine organisms. *Science*, 261, 78-82.

Caron, F, d'Errico, F, Del Moral, P, Santos, F & Zilhão, J (2011) The Reality of Neanderthal Symbolic Behaviour at the Grotte du Renne, Arcy-sur-Cure, France. *PlosOne,* 6 (6), e21545.

Caronni, S, Cristo, B & Torelli, A (2007) Tentativi di reimpianto del mollusco bivalve *Pinna nobilis* (Linneo, 1758) in una AMP della Sardegna. *Biologia Marina Mediterranea*, 14, 98-99.

Cecere, E & Mellea, S (Ed.) (2009) *Frammenti di mare. Taranto e l'antica molluschicoltura*. Fondazione Michelagnoli e CNR-IAMC, Taranto. 1-252.

Cerruti, A (1937) Primi esperimenti d'allevamento della *Pinna nobilis* L. nel Mar Piccolo di Taranto. *La ricerca scientifica*, Roma, 2 (1), 7-8.

Cerruti, A (1939) Ulteriori notizie sull'allevamento della *Pinna nobilis* L. nel Mar Piccolo di Taranto. *La ricerca scientifica*, Rome, 18, 1110-1121.

Cesari, P & Pellizzato, M (1985) Molluschi pervenuti in Laguna di Venezia per apporti volontari o casuali. Acclimazione di *Saccostrea commercialis* (Iredale & Roughley, 1933) e di *Tapes philippinarum* (Adams & Reeve, 1850). *Bollettino Malacologico, Milan*, 21 (10-12), 237-274.

Cianfarelli, S (2009) I molluschi della Provincia di Pistoia, le specie da tutelare e quelle da combattere. *Quaderni del Padule di Fucecchio*, 6, 1-112.

Civeyrel, L & Simberloff, D (1996) A tale of two snails, is the cure worse than the disease? *Biodiversity and Conservation,* 5 (10), 1231-1252.

Colonese, AC, Mannino MA, Bar-Yosef Mayer, DE, Fa, DA, Finlayson, JC, Lubell, D & Stiner, MC (2010) Marine mollusc exploitation in Mediterranean prehistory. An overview. *Quaternary International,* 239: 86-103.

Consorzio Ostricoltori Spezia (1914) *Istruzioni per preparare, conservare e cuocere le ostriche e i mitili*. La Spezia. 1-24.

Cortese, GE, Del Lucchese, A & Garibaldi, P (2004) *Charonia* sp., uno strumento musicale del Neolitico? *Preistoria alpina*, Trento, 40, 91-96.

Cosgrove, JA (1987) Aspects of the natural history of *Octopus dofleini*, the giant pacific octopus. M.Sc. Thesis. *Department of Biology, University of Victoria (Canada)*. 1-101.

Cossignani, T & Poppe, GT (1994) *Conchiglie ed Arte primitiva.* L'Informatore Piceno Ed, Ancona.

Cossignani, T (1992) *Cammei, gioielli dalle conchiglie.* L'Informatore Piceno Ed, Ancona.

Cossignani, T (2002) *Conchiglie, Maschere e Misteri.* L'Informatore Piceno Ed, Ancona.

Cossignani, T (2003) *Conchiglie, perle e madreperle.* L'Informatore Piceno Ed, Ancona.

Cossignani, T (2005) *Africa, misteri e conchiglie.* L'Informatore Piceno Ed, Ancona.

Costa Neto, EM (2006) Os moluscos na zooterapia, medicina tradicional e importância clínico-farmacológica. *Biotemas*, 19 (3), 71-78.

Coste, V (1861) *Voyage de exploration sur le littoral de la France et de l'Italie.* Imprimerie Imperial, Paris.

Cox, I (Ed) (1957) *The Scallop. Studies of a shell and its influences on humankind.* Shell Transport and Trading Company. 1-135.

Critchley, P (1975) *The art of Shellcraft.* Praeger Publ., New York. 1-96.

Crocetta, F, Bazzana A & Renda, W (2010) The infralittoral mobile bottom molluscs of Gizzeria harbour (Catanzaro, Italy). *Marine Biodiversity Records*, 3, 15.

Curiel, D, Bellemo, G., Marzocchi, M, Scattolin, M & Parisi, G (1998) Distribution of introduced Japanese macroalgae *Undaria pinnatifida*, *Sargassum muticum* (Phaeophyta) and *Antithamnion pectinatum* (Rhodophyta) in the Lagoon of Venice. *Hydrobiologia*, 385, 17-22.

Dance, SP (1986) *A history of shell collecting*. E.J. Brill, Leiden. 1-265.

De Panafieu, JB (2008) Le *Bestiaire marin. Histoires et legends des animaux des mers*. Plume de Carotte Ed, Toulouse. 1-170.

Del Vais, C (2006) *In piscosissimo mari. Il mare e le sue risorse tra antichità e tradizione*. Cabras (Oristano). 1-72.

Delluc, B & Delluc, G (1993). *Conoscere Lascaux*. Sud-Ouest Ed, Bordeaux. 1-64.

Dharmaraj, S, Shanmugasundaraman, K & Suja, CP (2004) Larval rearing and spat production of the windowpane shell *Placuna placenta*. *Aquaculture Asia*, 9, 20-28.

Dias, TLP, Leo Neto NA, & Alves RRN (2011) Molluscs in the marine curio and souvenir trade in NE Brazil: species composition and implications for their conservation and management. *Biodiversity and Conservation*, 20 (11), 2393-2405.

Doneddu, M & Trainito, E (2010) *Conchiglie del Mediterraneo*. Il Castello Ed. 1-272.

Doumange, F & Toulemont, A (Ed), (1992) Nacres et Perles. *Bulletin Musèe Oceanographique Monaco*, 8 (special), 1-162.

Ebisch, IMW, Thomas, CMG, Peters, WHM, Braat, DDM & Steegers-Theunissen, RPM (2007) The importance of folate, zinc and antioxidants in the pathogenesis and prevention of subfertility. *Human Reproduction Update*, 3 (2), 163-174.

Eldredge, N & Gould, SJ (1972) Punctuated equilibria, an alternative to phyletic gradualism. In: TJM Schopf (Ed), *Models in Paleobiology*. Freeman, Cooper and Co., San Francisco. 82-115.

Ellis, R (1999) *The search for the giant squid*. Penguin Books. 1-315.

FAO, (2005) *Review of the state of world marine fishery resources*. Fisheries Technical Paper, 457. Rome.

Farn, AE (1986) *Pearls, natural, cultured and imitation*. Butterworths gem books. 1-150.

Fenchel, T (1975) Character displacement and coexistence in mud snails (Hydrobiidae). *Oecologia*, 20, 19-32.

Fernàndez-Armesto, F (2002) *Near a Thousand Tables: A History of Food*. The Free Press, New York.

Figuier, L (1866) *La vie et les moeurs des animaux. Zoophytes et Mollusques*. Hachette, Paris, 1-500.

Fleming, G (1829) *Filosofia zoologica, ossia prospetto generale della struttura, funzioni e classificazione degli animali*. Pavia.

Ford, SE & Chintala, MM (2006) Northward expansion of a marine parasite, Testing the role of temperature adaptation. *Journal Experimental Marine Biology Ecology,* 339 (2), 226-235.

Galil, BS (2007) Loss or gain? Invasive aliens and biodiversity in the Mediterranean Sea. *Marine Pollution Bulletin,* **55**: 314-322.

Gangemi, A (2012) *Pericoli da vita acquatica*. Laruffa Ed, Reggio Calabria. 1-156.

Garaventa, F, Centanni, E, Pellizzato, F, Faimali, M, Terlizzi A & Pavoni, A (2007) Imposex and accumulation of organotin compounds in populations of *Hexaplex trunculus* (Gastropoda, Muricidae) from the Lagoon of Venice (Italy) and Istrian Coast (Croatia). *Marine Pollution Bulletin*, 54 (5), 615-622.

Gay, L (1988) *Ostrica, mare vivo*. Idea Libri, Milan. 1-96.

Giacobbe, S, Caltabiano M & Pugliesi, M (2009) The pelorias shell in the ancient coins, taxonomic attribution and implication in the management of Capo Peloro and Laghi di Ganzirri. *Biologia Marina Mediterranea*, 16 (1), 134-135.

Girod, A (2001) Le conchiglie. Studio malacologico. In: Bulgarelli, F, Gardini A & Melli, P (Eds), *Archeologia dei pellegrinaggi in Liguria*. M. Sabatelli Ed Savona. 1-160.

Gofas, S & Zenetos, A (2003) Exotic molluscs in the Mediterranean basin, current status and perspectives. *Oceanography and Marine Biology, an Annual Review*, 41, 237-277.

Gössling, S, Kunkel, T, Schumacher, K, & Zilger, M (2004) Use of molluscs, fish, and other marine taxa by tourism in Zanzibar, Tanzania. *Biodiversity and Conservation*, 13 (14), 2623-2639.

Graf, A (2002) *Miti, leggende e superstizioni del Medioevo*. Bruno Mondadori Ed, Milan.

Grassia, L & Vilboux, F (2008) *Ostriche passioni divine*. Sagep Ed, Genoa. 1-207.

Grimberg, C (1968) *L'alba della civiltà, Asia, Africa, Mediterraneo*. Dall'Oglio Ed.

Groeben, C & Gambi, MC (1992) *Lo Guarracino che jeva pe mare.* Gaetano Macchiaroli Ed, Naples. 1-93.

Guidetti, P & Boero, F (2004) Desertification of Mediterranean rocky reefs caused by date-mussel, *Lithophaga lithophaga* (Mollusca, Bivalvia) fishery, effects on adult and juvenile abundance of a temperate fish. *Marine Pollution Bullettin*, 48, 978-982.

Haddad, V, de Paula Neto, JB & Cobo, VJ (2006) Venomous mollusks, the risks of human accidents by *Conus* snails (Gastropoda, Conidae) in Brazil. *Revista da Sociedade Brasileira de Medicina Tropical*, 39, 498-500.

Hall-Spencer, JM, Rodolfo-Metalpa R, Martin, S, Ransome, E, Fine, M, Turner, SM, Rowley, SJ, Tedesco, D & Buia, MC (2008) Volcanic carbon dioxide vents show ecosystem effects of ocean acidification. *Nature,* 454: 96-99.

Haszprunar, G (2001) Mollusca. *Encyclopedia of Life Sciences*. John Wiley & Sons, Ltd.

Henshilwood, CS, d'Errico F & Watts, I (2009) Engraved ochres from the Middle Stone Age levels at Blombos Cave, South Africa. *Journal Human Evolution,* **57:** 27-47.

Henshilwood, CS, d'Errico, F, van Niekerk, KL, Coquinot, Y, Jacobs, Z, Lauritzen, SE, Menu, M & García-Moreno, R (2011) A 100,000-year-old ochre-processing workshop at Blombos Cave, South Africa. *Science,* 334*,* 219-222.

Hodgkin, AL, Huxley, AF & Katz, B (1952) Measurement of current-voltage relations in the membrane of the giant axon of *Loligo. Journal of Physiology,* 116, 424-448.

Hoffmann, E, Ford, S, Powell, E & Klinch, J (2001) Modeling studies of the effect of climate variability on MSX disease in eastern oyster (*Crassostrea virginica*) populations. *Hydrobiologia*, 460, 195-212.

Hogendorn, J & Johnson, M (Eds) (2003) *The shell money of the slave trade.* African Studies Centre, Cambridge.

Houart, R (2001) *A review of the recent Mediterranean and Northeastern Atlantic species of Muricidae*. Evolver Ed, Rome.

Isolani, B & Manachini, B (1996) *Le conchiglie di Lamarck.* Consorzio per il Centro Interuniversitario di Biologia Marina "G. Bacci" Livorno. 1-63.

Issel, A (1882) *Istruzioni pratiche per l'ostricoltura e la mitilicoltura*. Tipografia del Regio Istituto dei Sordo Muti, Genoa. 1-260.

Issel, A (1908) Liguria Preistorica. *Atti della Società Ligure di Storia Patria,* Genoa*,* 40*,* 41-92.

Issel, R (1918) *Biologia marina.* Hoepli Ed, Milan. 1-607.

Jackson, JW (1917) *Shells as evidence of the migrations of early culture.* Manchester.

Joordens, CA, d'Errico, F, Wesselingh, FP, Munro S de Vos, J, Wallinga, J, Ankjærgaard, C, Reimann, T, Wijbrans, JR, Kuiper, KF, Mücher, HJ, Coqueugniot, H, Prié, HV, Joosten, I, van Os, B, Schulp, AS, Panuel M., van der Haas V, Lustenhouwer W, Reijmer, JJG & Roebroeks, W (2015) *Homo erectus* at Trinil on Java used shells for tool production and engraving. *Nature* 518, 228-231.

Jorgensen, CB (1990) *Bivalve Filter Feeding*. Olsen & Olsen Ed. 1-140.

Kandel, ER (1979) *Behavioral biology of Aplysia*. W.H. Freeman Ed, San Francisco. 1-463.

Karali, L (1999) *Shells in Aegean Prehistory*. British Archaeological Reports, International Series, Oxford. 1-761.

Kessler, CG & Russell, HD (1978) Leopold and Rudolph Blaschka's nudibranch glass models. *The Nautilus*, 92 (4), 167-172.

Kubodera, T & Mori, K (2005) First-ever observation of a live giant squid in the wild. *Proceedings of the Royal Society B, Biological Sciences,* 272 (1581), 2583-2586.

Kurlansky, M (2006) *The Big Oyster, History on the half shell*. Jonathan Cape Ed, 1-320.

Lalli, CM & Gilmer, RW (1989) *Pelagic snails. The biology of the holoplanktonic Gastropod Mollusks*. Stanford University Press, California. 1-259.

Lefkaditou, E, Corsini-Foka, M & Kondilatos, G (2009) Description of the first Lessepsian squid migrant, *Sepioteuthis lessoniana* (Cephalopoda, Loliginidae), in the Aegean Sea (Eastern Mediterranean). *Mediterranean Marine Science,* 10 (2), 87-97.

Leiva, GE & Castilla, JC (2002) A review of the world marine gastropod fishery, evolution of the catches, management and the Chilean experience. *Revue Fish Biology Fisheries,* 11, 283-300.

Leo Neto, NA & Alves, RRN (2010) A Natureza sagrada do Candomblé: análise da construção mística acerca da Natureza em terreiros de candomblé no Nordeste de Brasil. *Interciencia,* 35 (8), 568-574.

Leo Neto, NA, Mourão, JS & Alves, RRN (2011) "It all begins with the head": initiation rituals and the symbolic conceptions of animals in Candomblé. *Journal of Ethnobiology,* 31 (2), 244–261.

Leo Neto, NA, Voeks, RA, Dias, TLP & Alves, RRN (2012) Mollusks of Candomble: symbolic and ritualistic importance. *Journal of Ethnobiology and Ethnomedicine* 8 (1),10.

Leonhard, K (2007) Shell collecting. On 17th century conchology, curiosity cabinets and still life painting. In: Enenkel, KAE & Smith PJ (Eds), *Early modern zoology, the construction of animals in science*. Brill, Leiden.

Leroi-Gourhan, A (1961) Les fouilles d'Arcy-sur-Cure (Yonne). *Gallia préhistoire,* 4, 3-16.

Levri, EP, Kelly, AA & Love, E (2007) The invasive New Zealand mud snail (*Potamopyrgus antipodarum*) in Lake Erie. *Journal of Great Lakes Research*, 33, 1-6.

Lori, E & Cianfarelli, S (2006) New records of *Dreissena polymorpha* (Pallas, 1771) (Mollusca, Bivalvia, Dreissenidae) from Central Italy. *Aquatic Invasions*, 1 (4), 281-283.

MacColl, R, Galivan, J, Berns, DS, Nimec, Z, Guard-Friar, D & Wagoner, D (1990) The chromophore and polypeptide composition of *Aplysia* ink. *Biological Bulletin*, 179, 326-331.

Mann, R, Harding, J & Westcott, E (2006) Occurrence of imposex and seasonal patterns of gametogenesis in the invading rapa whelk *Rapana venosa* from Chesapeake Bay, USA. *Marine Ecology Progress Series,* 310, 129-138.

Mannino, MA & Thomas, KD (2002) Depletion of a resource? The impact of prehistoric human foraging on intertidal mollusc communities and its significance for human settlement, mobility, and dispersal. *World Archaeology,* 33 (3), 452-474.

Mazzola, A (1992) Le specie alloctone e l'acquacoltura. In: Protezione della fauna marina e introduzione di specie alloctone. *Bollettino dei Musei e degli Istituti Biologici dell'Università di Genova*, 56-57, 225-234.

Meinhardt, H (1995) *The algorithmic beauty of sea shell*. Springer-Verlag. 1-204.

Mills, EL, Rosenberg, G, Spidle, AP, Ludyanskiy, M, Pligin, Y & May, B (1996) A review of the biology and ecology of the quagga mussel (*Dreissena bugen-*

sis), a second species of freshwater Dreissenid introduced to North America. *American Zoologist*, 36, 271-286.

Mizzan, L, Trabucco, R & Tagliapietra, P (2005) Nuovi dati sulla presenza e distribuzione di specie alloctone del macrozoobenthos della Laguna di Venezia. *Bollettino Museo civico Storia Naturale di Venezia*, 56, 69-88.

Morazzoni, G & Pasquato, M (1953) *Le conterie veneziane*. Società Veneziana Conterie e Cristallerie Ed, Venice.

Morelli, P & Saulle, S (1998) *Anna Comnena, la poetessa epica (1083-1153)*. Jaca Book.

Mottana, A (1999) Oggetti e concetti inerenti le Scienze Mineralogiche ne *La composizione del mondo con le sue cascioni* di Restoro d'Arezzo (anno 1282). *Rendiconti Lincei: Scienze Fisiche e Naturali,* s. 9, 10, 133-229.

Nardi, G & Braccia, A (2004) Prima segnalazione di *Corbicula fluminea* (OF Müller, 1774) per il Lago di Garda (provincia di Brescia) (Mollusca, Bivalvia, Corbiculidae). *Bollettino Malacologico,Milan*, 39 (9-12), 181-184.

Neal, B 2007. *Chinese mother-of-pearl gaming counters*. Chezbill Ed, 1-38.

Niero, I (2003) Sulla presenza in Veneto e Centro Italia di *Anodonta woodiana woodiana* (Lea, 1834) (Mollusca, Bivalvia). *Bollettino Museo civico Storia Naturale di Venezia*, 54, 29-33.

Occhipinti Ambrogi, A & Galil, B (2010) Marine alien species as an aspect of global change. *Advances in Oceanography and Limnology*, 1 (1), 199-218.

Pettit, GR (1997) Dolastatins. In: *Progress in the Chemistry of Organic Natural Products*. Springer Verlag, Wien.

Pianka, ER (1994) *Evolutionary Ecology*. Harper Collins Publ, 1-486.

Piersanti, C (1926) *I molluschi e le conchiglie*. Manuali Hoepli, 1-502.

Pizzoni, F (1997) *Il giardino, arte e storia*. Leonardo Arte Ed, Milan, 1-263.

Polanyi, K (1980) *Economie primitive, arcaiche e moderne*. Einaudi Ed, Turin, 1-340.

Polanyi, K (1987) *Il Dahomey e la tratta degli schiavi*. Einaudi Ed, Turin, 1-265.

Poncet, J (1999) The dolastatins, a family of promising antineoplastic agents. *Current Pharmaceutical Design*, 5, 139.

Poutiers, JM (1993) Les coquillages comestibles de France. In: Elzière-Papayanni P (Ed), *Coquillages*. Informations Techniques des Servies Vétérinaires Français, Paris, 13-102.

Pozzoli, G, Noel, F-J M, Romani F & Peracchi, A (1825) *Dizionario d'ogni mitologia e antichità*. 6. Fanfari Ed.

Pranovi, F & Giovanardi, O (1998) La pesca di molluschi bivalvi nella Laguna di Venezia, effetti e conseguenze. *Biologia Marina Mediterranea*, 2, 121-122.

Raybaudi, L (1979) *Cypraea (Leporycypraea) valentia*. *La Conchiglia*, Rome, 11 (126-127), 11-14.

Rayment, WJ (2008) *Crepidula fornicata*. Slipper limpet. *Marine Life Information Network, Biology, and Sensitivity Key Information Sub-programme*. Marine Biological Association of the United Kingdom, Plymouth.

Reese, DS (1991) The trade of Indo-Pacific shells into the Mediterranean Basin and Europe. *Oxford Journal of Archeology*, 10 (2), 159-196.

Ripa, C (1593) *Iconologia overo descrittione dell'imagini universali cavate dall'antichità et da altri luoghi da Cesare Ripa Perugino*. Giovanni Osmarino Gigliotti Stampatore con Privilegio et con Licenza de' Superiori.

Ristine, JD (2009) *Clams, Oysters, & Scallops*. Schiffer Publ, 1-128.

Rivière, E (1887) *De l'antiquitè de l'Homme dans les Alpes Maritimes*. Paris.

Roberto, U (2005) *Ioannis Antiocheni Fragmenta ex Historia Chronica*. W. de Gruyter & Co (Ed).

Rogers, JE (1908) *The Shell book*. Doubleday, Page Co, New York.

Roghi, G (1963) Le conchiglie moneta. *Mondo Sommerso*, 10.

Rondelet, G (1554) *De Piscibus marinis*. Lugdini.

Roper, CFE, Sweeney, MJ & Nauen, CF (1984) *Cephalopods of the world. An annotated of species of interest to fisheries.* FAO Species Catalogue, 3. Rome.

Runham, NW (1991) Commercial farming of *Achatina* for food in the U.K. In: *Proceeding Tenth International Malacological Congress*, Tubingen, 1, 295-296.

Russo, GF & Cicogna, F (1992) Il dattero di mare, *Lithophaga lithophaga*, e gli effetti distruttivi della sua pesca sull'ambiente marino costiero, problemi e prospettive. *Bollettino dei Musei e degli Istituti Biologici dell'Università di Genova*, 56-57, 165-193.

Russo, GF (1999) Cambiamenti faunistici nel Mar Mediterraneo, un approccio zooarcheologico. In: *Mediterraneo, condizioni e sviluppi del cambiamento. Quaderni Accademia Internazionale di Scienze e Tecniche Subacquee*, Ustica, 22, 63-73.

Sachs, C (1940) *The History of Musical Instruments*. New York.

Safriel, UN, Erez, N & Keasar, T (1993) Long term effects of transplanted limpets on an experimental marine biofouling community. *Biofouling*, 6 (3), 261-278.

Sala, N & Cappellato, G (2003) *Viaggio matematico nell'arte e nell'architettura*. Franco Angeli Ed, 1-256.

Saul, M (1974) *Shells, an illustrated guide to a timeless and fascinating world.* Hamlyn Ed., London. 1-192.

Savini, D, Occhipinti-Ambrogi, A & Castellazzi, M (2007) Distribution of the alien gastropod *Rapana venosa* in the northern Adriatic Sea. *Rapports de la Commission Internationale pour l'Exploration Scientifique de la Mer Méditerranée,* 38, 590.

Scarpa, A (1980) *Etnomedicina*. Franco Lucisano Ed, Milan.

Schapira, D, Montano, IA, Antczak, A & Posada, JM (2009) Using shell middens to assess effects of fishing on queen conch (*Strombus gigas*) populations in Los Roques Arcipelago National Park, Venezuela. *Marine Biology*, 156, 787-795.

Schloesser, DW & Nalepa, TF (1994) Dramatic decline of unionid bivalves in offshore waters of Western Lake Erie after infestation by the zebra mussel, *Dreissena polymorpha. Canadian Journal of Fisheries and Aquatic Sciences*, 51 (10), 2234-2242

Scotti, G & Chemello, R (2000) I molluschi marini mediterranei degni di protezione, stato di conoscenza e forme di tutela. In: Chemello, R & Manganelli, G (Eds), *La conservazione della natura e dei molluschi. Atti del 1° Workshop Internazionale di Malacologia, Bollettino Malacologico, Milan*, 36, 61-70.

Seba, A (1734-65) *Locupletissimi rerum naturalium thesauri accurata descriptio, et iconibus artificiosissimis expressio, per universam physices historiam. Opus, cui, in hoc rerum genere, nullum par exstitit, ex toto terrarum orbe collegit, digessit, descripsit, et depingendum curavit Albertus Seba.* Waesberg, J, Wetstenium, J & Smith, G, Amsterdam.

Sharabati, D (1981) *Saudi Arabian Seashells*. B.V. Weert Ed, 1-119.

Sheumack, DD, Howden, ME, Spence, I & Quinn, RJ (1978) Maculotoxin, a neurotoxin from the venom glands of the octopus *Hapalochlaena maculosa* identified as tetrodotoxin. *Science,* 199 (4325), 188-189.

Singer, RT (1999) *Edo, Art in Japan 1615-1868.* Yale University Press, New Haven. 1-480.

Skeates, P (1991) Triton's trumpet, a neolithic symbol in Italy. *Oxford Journal of Archeology*, 10, 17-31.

Southward, AJ, Southward, EC, Dando, JA, Hugues, JA, Kennicutt, MC, Alcalà-Herrera, J & Leahy, Y (1997) Behaviour and feeding of the nassariid gastropod *Cyclope neritea*, abundant at hydrothermal brine seeps off Milos (Aegean Sea). *Journal Marine Biological Association UK,* 77, 753-771.

Spencer, BE (Ed) (2002) *Molluscan Shellfish Farming*. Blackwell Publishing, Oxford, UK. 1-274.

Stearns, REC (1889) Ethno-conchology: a study of primitive money. *Annual Reports US Natural Museum* (1887), 297-334.

Steneck, RS & Carlton, JT (2000) Human alterations of marine communities, students beware! In: Berness, MD, Gaines SD & Hay, ME (Eds), *Marine Community Ecology*. Sinauer Ass., Massachussets. 445- 468.

Stiner, MC, Pehlevan, C, Sagir, M & Özer, I (2002) Zooarchaeological studies at Üçagzl Cave, preliminary results on Paleolithic subsistence and shell ornaments. *Kazı Sonuçları Toplantıları*, Ankara, 17, 29-36.

Stott, R (2004) *Oyster*. Reaktion Books Ltd, London. 1-240.

Strom, A, Francis, RC, Mantua, NJ, Miles, EL & Peterson, DL (2004) North Pacific climate recorded in growth rings of geoduck clams, a new tool for paleoenvironmental reconstruction. *Geophysical Research Letters*, 31, L06206.

Sutty, L (1984) *Seashell treasures of the Caribbean*. Dutton Ed, New York. 1-128.

Taburiaux, J (1985) *Pearls, their origin, treatment, and identification*. N.A.G. Press, Ipswich, Suffolk. 1-247.

Terlizzi, A, Delos AL, Garaventa, F, Faimali, M &. Geraci, S (2004) Limited effectiveness of Marine Protected Areas, imposex in *Hexaplex trunculus* (Gastropoda, Muricidae) populations from Italian marine reserves. *Marine Pollution Bulletin*, 48, 186-190.

Thomas, I (2007) *The shell, a world of decoration and ornament*. Thames Hudson, London, 1-256.

Thompson, d'Arcy W (1961) *On growth and form*. (Abridged Ed.). Cambridge University Press, 1-346.

Thompson, TE & Bennett, I (1969) *Physalia* nematocysts, utilized by mollusks for defense. *Science,* 166, 1532-1533.

Tiberi, N (1879) Le conchiglie Pompeiane. *Bollettino Società Malacologica Italiana*, 5, 139-151.

Torchio, M (1983) Osservazioni e considerazioni sulle origini della malacologia (nota preliminare). *Quaderni Civica Stazione Idrobiologica*, Milan, 11, 11-40.

Travers, LA (1962) *The romance of shells, in nature and art*. Barrows Ed.

Verrill, AH (1939) *Strange Sea Shells and their stories. How they are made and grow. How they are colored and the patterns produced*. LC Page & Company, 1-220.

Viviano, G (2008) *Babbaluci* come elementi rituali nel festino di Santa Rosalia a Palermo. *Notiziario Società Italiana Malacologia,* 26 (9-12), 8-10.

Voultsiadou, E, Koutsoubas, D & Achparaki, M (2010) Bivalve mollusc exploitation in Mediterranean littoral communities, a historical approach. *Journal of Biological Research*, Thessaloniki, 13, 35-45.

Voultsiadou, E (2010) Therapeutic properties and uses of marine invertebrates in the ancient Greek world and early Byzantium. *Journal of Ethnopharmacology*, 130, 237-247.

White, AW (1977) Dinoflagellate toxins as probable cause of an Atlantic herring (*Clupea harengus harengus*) kill, and pteropods as apparent vector. *Journal Fish Research Bd. Canada*, 34, 2421-2424.

Whiteford, AH (2002) *North American Indian Arts*. St. Martin's Press, New York, 1-160.

Williams, W (2010) *Kraken*. Abrams, New York, 1-223.

Williamson, JA, Fenner, PJ Burnett, JW & Rifkin, J (1996) *Venomous and poisonous Marine Animals, a medical and biological handbook*. UNSW Press, 65-68.

Wilson, EO (1992) *The Diversity of Life*, Harvard University Press, 1-589.

Wood, JB (1931) Illinois, crossroads of the continent. *National Geographic Magazine*, 59 (5), 523-594.

Wood, E & Wells, S (1995) The shell trade: a case for sustainable utilization. In: Kay EA (Ed) *The conservation biology of molluscs*. IUCN Species Survival Commission, Cambridge.

Wurtz, M (1989) *Nautilus*. Oceanographic Museum of Monaco. 1-43.

Yonge, CM (1930) *A year on the Great Barrier Reef, the story of corals and of the greatest of their creations.* Putnam Ed, London.

Yonge, CM (1960) *Oysters*. Collins Eds, London, 1-290.

Zenetos, A, Konstantinou, F & Konstantinou, G (2009) Towards homogenization of the levantine alien biota, additions to the alien molluscan fauna along the Cypriot coast. *Marine Biological Association of the United Kingdom, Marine Biodiversity Records*, 2, 1-7.

Zenetos, A, Çinar, ME, Pancucci-Papadopoulou, MA, Harmelin, JG, Furnari, G, Andaloro, F, Bellou, N, Streftaris, N & Zibrowius, H (2005) Annotated list of marine alien species in the Mediterranean with records of the worst invasive species. *Marine Science*, 6 (2), 63-118.

Zilhão, J, Angelucci, DE, Badal-García, E, d'Errico, F, Daniel, F, Dayet, L, Douka, K, Higham, TFG, Martínez-Sánchez, M-J, Montes-Bernárdez, R, Murcia-Mascarós, S, Pérez-Sirvent, C, Roldán-García, C, Vanhaeren, M, Villaverde, V, Wood, R & Zapata, J (2010) Symbolic use of marine shells and mineral pigments by Iberian Neanderthals. *Proceedings National Academy Sciences USA,* 107 (3), 1023-1028.

Figure 110: Snails crawling over the *Sagrada Familia* in Barcelona.

Subject Index

C

D

H

I

J

K

L

M

N

O

P

Q

R

U

V

W

X

Y

Z

Made in the USA
Las Vegas, NV
03 November 2026

57422844R00214